THE STRUCTURE AND DYNAMICS OF CITIES

With over half of the world's population now living in urban areas, the ability to model and understand the structure and dynamics of cities is becoming increasingly valuable. Combining new data with tools and concepts from statistical physics and urban economics, this book presents a modern and interdisciplinary perspective on cities and urban systems. Both empirical observations and theoretical approaches are critically reviewed, with particular emphasis placed on derivations of classical models and results, along with analysis of their limits and validity. Key aspects of cities are thoroughly analyzed, including mobility patterns, the impact of multimodality, the coupling between different transportation modes, the evolution of infrastructure networks, spatial and social organization, and interactions between cities. Drawing upon knowledge and methods from areas of mathematics, physics, economics, and geography, the resulting quantitative description of cities will be of interest to all those studying and researching how to model these complex systems.

MARC BARTHELEMY is a senior researcher at the Institute of Theoretical Physics (IPhT-CEA/CNRS) in Saclay and a member of the Center of Social Analysis and Mathematics (CAMS-EHESS/CNRS). He has worked on applications of statistical physics to complex networks, epidemiology, and more recently on spatial networks, as well as co-authoring *Dynamical Processes on Complex Networks* with Alain Barrat and Alessandro Vespignani. Focusing on both data analysis and modeling, he is currently working on various aspects of the emerging science of cities.

THE STRUCTURE AND DYNAMICS OF CITIES

Urban Data Analysis and Theoretical Modeling

MARC BARTHELEMY

Institut de Physique Théorique
Commissariat à l'Energie Atomique

and

Centre d'Analyse et de Mathématique Sociales
École des Hautes Études en Sciences Sociales

CAMBRIDGE
UNIVERSITY PRESS

University Printing House, Cambridge CB2 8BS, United Kingdom

Cambridge University Press is part of the University of Cambridge.

It furthers the University's mission by disseminating knowledge in the pursuit of
education, learning and research at the highest international levels of excellence.

www.cambridge.org
Information on this title: www.cambridge.org/9781107109179

© Cambridge University Press 2016

This publication is in copyright. Subject to statutory exception
and to the provisions of relevant collective licensing agreements,
no reproduction of any part may take place without the written
permission of Cambridge University Press.

First published 2016

Printed in the United Kingdom by TJ International Ltd. Padstow Cornwall

A catalogue record for this publication is available from the British Library

Library of Congress Cataloguing in Publication data
Names: Barthelemy, Marc, 1965– author
Title: The structure and dynamics of cities : urban data analysis and
theoretical modeling / Marc Barthelemy, Institut de Physique Théeorique
Commissariat à l'Energie Atomique and Centre d'Analyse et de
Mathématique Sociales Ecole des Hautes Etudes en Sciences Sociales.
Description: New York : Cambridge University Press, 2016. |
Includes bibliographical references and index.
Identifiers: LCCN 2016021104 | ISBN 9781107109179 (hardback)
Subjects: LCSH: Cities and towns–Statistical methods. |
Quantitative research. | System analysis.
Classification: LCC HT110 .B37 2016 | DDC 307.76072/7–dc23
LC record available at https://lccn.loc.gov/2016021104

ISBN 978-1-107-10917-9 Hardback

Cambridge University Press has no responsibility for the persistence or accuracy of
URLs for external or third-party internet websites referred to in this publication,
and does not guarantee that any content on such websites is, or will remain,
accurate or appropriate.

To Esther and Rebecca

Contents

Preface		*page* xi
Acknowledgments		xvii
1	**Urban systems**	1
	1.1 A science of cities	2
	1.1.1 The nature of the problem	2
	1.1.2 What is a city? Origins and definitions	4
	1.2 Spatial and temporal scales	8
	1.2.1 Population	8
	1.2.2 Area, density, and volume of cities	12
	1.2.3 Time scales	19
	1.3 Naïve scaling	20
	1.3.1 Surface area	21
	1.3.2 Total length of roads	21
	1.3.3 Total daily commuting distance	23
2	**Models and methods**	25
	2.1 Statistical physics of complex systems	25
	2.2 The shape of a science of cities	26
	2.3 How many parameters?	28
	2.3.1 Statistical physics and relevant parameters	28
	2.3.2 Modeling cities	29
	2.4 Critiques of urban economics	31
	2.4.1 Interactions and equilibrium	32
	2.4.2 Invariance with respect to utility choice	33
	2.5 Data	37
	2.5.1 Sources	38
	2.5.2 Different types of data	39
	2.5.3 Data are not enough: models	43
	2.6 The barriers to interdisciplinarity	45

3 The spatial organization of cities — 47

- 3.1 Optimal locations — 47
  - 3.1.1 Distribution of public facilities — 47
  - 3.1.2 Distribution of retail stores — 49
- 3.2 Measuring a polycentric structure — 52
  - 3.2.1 Definition — 52
  - 3.2.2 Identifying and counting hotspots — 54
- 3.3 Polycentricity: Classical approaches — 57
  - 3.3.1 The Fujita–Ogawa model — 57
  - 3.3.2 The edge-city model — 63
- 3.4 Revisiting the Fujita–Ogawa model — 65
  - 3.4.1 A complex quantity described as random — 65
  - 3.4.2 Monocentric-polycentric transition — 67
  - 3.4.3 Number of centers — 68
  - 3.4.4 Consequences for mobility — 70
  - 3.4.5 CO_2 emission and gasoline consumption — 72
  - 3.4.6 Urban villages — 75
  - 3.4.7 The most economical population distribution — 76

4 Infrastructure networks — 78

- 4.1 Roads and streets: patterns — 78
  - 4.1.1 Length of the network — 79
  - 4.1.2 Statistics of blocks — 81
  - 4.1.3 Structure of paths — 89
- 4.2 Evolution of the road network — 95
  - 4.2.1 Basic properties — 96
  - 4.2.2 Simplicity profile — 99
  - 4.2.3 Betweenness centrality impact — 101
  - 4.2.4 Evolving patterns of betweenness centrality — 102
  - 4.2.5 Modeling the road network — 104
- 4.3 Subways — 111
  - 4.3.1 All large cities have a subway system — 111
  - 4.3.2 Convergence to a universal structure — 112
  - 4.3.3 Scaling and modeling for subways — 120
- 4.4 Digression: Railroads — 125
  - 4.4.1 Scaling — 125
  - 4.4.2 Are subways and railroads the same? — 127

5 Mobility patterns — 129

- 5.1 Typology of origin–destination matrices — 130
  - 5.1.1 Extracting coarse-grained information from OD matrices — 132
  - 5.1.2 Comparing mobility networks — 134

			138
	5.2	Modeling mobility patterns	138
		5.2.1 Statistics of flows: from gravity to radiation	138
		5.2.2 Commuting and income	142
	5.3	Human mobility: Levy flights or accelerated walkers?	151
		5.3.1 Back to basics: empirical observations	152
		5.3.2 Modeling the hierarchy of modes	155
6	**Multimodality in cities**		161
	6.1	A multilayer network view of urban navigation	162
		6.1.1 Empirical observations of multimodality	162
		6.1.2 Characterizing the multilayer system	165
	6.2	The effect of coupling	172
		6.2.1 A toy model	173
		6.2.2 Optimal velocity for the road–subway system	177
	6.3	Information perspective on navigation in cities	185
		6.3.1 Simplest paths	186
		6.3.2 Information entropy	186
		6.3.3 Information threshold: 8 bits	188
		6.3.4 Effect of multimodal couplings	190
7	**Socioeconomic aspects**		193
	7.1	Classical models of urban economics	194
		7.1.1 Why discuss these models here?	194
		7.1.2 The Alonso–Muth–Mills model	194
		7.1.3 Beckmann's model: space and the social network	200
	7.2	Segregation and income structure of cities	203
		7.2.1 A null model for spatial segregation	204
		7.2.2 The emergent social stratification of cities	205
	7.3	Modeling segregation	210
		7.3.1 Transportation modes in the Alonso–Muth–Mills model	210
		7.3.2 A simple model for tie formation	213
		7.3.3 Statistical physics of the Schelling model	214
		7.3.4 Collective versus individual dynamics	217
	7.4	Scaling in urban systems	219
		7.4.1 What is scaling?	219
		7.4.2 Theoretical approaches	222
8	**Systems of cities**		225
	8.1	Population distribution	225
		8.1.1 The number of cities and the largest city	226
		8.1.2 Gibrat, Gabaix, and diffusion with noise	228

8.2	Central place theory and spatial fluctuations	236
	8.2.1 Outline of Christaller's theory	236
	8.2.2 Spatial fluctuations	237

9 Toward a new science of cities — 242
9.1 What is our "understanding"? — 242
9.2 Measuring the death and life of great cities — 244
9.3 The future of the city — 245
9.4 Concluding thoughts — 247

References — 248
Index — 261

Preface

Most of the world's people are now living in cities and urbanization is expected to keep increasing in the near future. The resulting challenges are complex, difficult to handle, and range from increasing dependence on energy, to the emergence of socio-spatial inequalities, to serious environmental and sustainability issues. Understanding and modeling the structure and evolution of cities is then more important than ever as policy makers are actively looking for new paradigms in urban and transport planning.

The recent advances obtained in the understanding of cities have generated increased attention to the potential implication of new theoretical models in agreement with data. Questions such as urban sprawl, effects of congestion, dominant mechanisms governing the spatial distribution of activities and residences, and the effect of new transportation infrastructures are fundamental questions that we need to understand if we want a harmonious development of cities in the future, from both social and economic points of view.

Cities were for a long time the subject of numerous studies in a large number of fields. Discussion of the ideal city can be traced back at least to the Renaissance, and more recently scientists have tried to describe quantitatively the formation and evolution of cities. Regional science and then quantitative geography addressed various problems such as the spatial organization of cities, the impact of infrastructures, and transport. It is remarkable to note that as early as the 1970s quantitative geographers realized the crucial importance of networks in these systems, and produced visionary studies about networks, their evolution, and the complexity of cities (Haggett et al. 1977).

These studies were further developed mathematically by economists who discussed the interplay between space and economic aspects in cities. Many important models find their origin in the seminal paper of Von Thunen and describe isolated, monocentric cities in terms of utility maximization subject to budget constraints. These models allowed spatial economics to get a grasp of the relations between space, income, and transportation; for example. Japanese economists Fujita and Ogawa discussed the impact of agglomeration effects between firms in a general model that deals with the location choice for individuals and companies. However, as is often the case in the physics of complex systems, writing complicated equations that are essentially impossible to solve does not lead to much progress in our understanding. In fact, at the heart of many economic processes such as location choice or job search lies the problem of how individuals make decisions. Describing this process is a rather formidable task and utility optimization or other optimal strategies are interesting attempts to accomplish it. Many theoretical papers and books discuss these various problems, but in most of these approaches there is one fundamental flaw: the relation between these theoretical models and the reality.

This is where a recent game changer enters the arena: huge amounts of data on every possible aspect of cities have suddenly (at the human scale) become available. In the hope of uncovering underlying laws governing the dynamics and the evolution of these systems, researchers can now begin the systematic analysis of many different cities. Today we are making the first transversal studies to try to uncover universal behavior, independent from the history, culture, or geography of these systems. This trend is reinforced by ideas such as the self-organization of complex systems. The view that a city can be described as a complex, self-organized system is comparatively recent and can be traced back to studies done by geographers such as Pumain in France and Batty in the UK, but had also been rapidly understood and accepted by leading economists such as Krugman.

In addition to geographers and economists, statistical physicists have recently become interested in cities. Indeed, the availability of data, together with the large size of cities, the large number of their constituents, and the variety of processes, are all ingredients that are both very attractive and challenging for a statistical physicist. Attractive, because statistical physics has a long history of understanding and characterizing emergent macroscopic phenomena in terms of the dynamical evolution of the basic constituents of the sytem. Also, the recent large-scale analysis of different cities has provided evidence for common emergent properties that go beyond specific features of these systems and thus lead to the possibility of proposing general models. Challenging, because individuals and institutions are not atoms, and because there are so many aspects to cities that it might be extremely difficult to disentangle the different processes and to reach a clear, minimal model in agreement with data.

We have thus witnessed in the last few years an explosion of new studies on cities with the help of new data and new tools to analyze them. Various models and aspects have recently been discussed from this perspective, revisiting urban economics in the light of statistical physics and usually informed by new datasets. Many results have been obtained and the purpose of this book is an attempt to provide some kind of unified view of some of these new models, their variants, and their impact on our understanding of cities. In doing so, we will make a special effort to define, for each model or phenomenon studied, the appropriate language used in the field, and to offer the reader a mapping between languages and techniques used in different disciplines.

We will discuss many aspects of cities and for each one we will essentially focus on new approaches that are not necessarily mainstream, but that are connected to data and that propose a new perspective on a phenomenon. Each chapter is as much as possible self-contained and addresses a specific phenomenon that takes place in cities. For the sake of clarity and readability we have tried to be as modular as possible, in order to allow the reader interested in just one phenomenon to focus essentially on the corresponding chapter.

The chapters are sequenced according to a rough measure of increasing complexity. In the first chapter, we discuss general features of cities and try to bring new perspectives whenever possible. We start with a general discussion of the definition of cities and what is currently believed about their origins. We then discuss important properties of cities – some of them known for a long time, such as the Zipf law for population – and their various spatial and temporal scales. In particular, we will also discuss the order of magnitude of areas, densities of cities, the number of cities, and the size of the largest city in each country. We end this first chapter with naïve scaling that relates various quantities among them. These elementary arguments are interesting as they provide some benchmarking for discussing empirical observations.

In the second chapter, we propose a discussion about methods to study cities. We discuss how statistical physics could help – from both a methodological and a conceptual point of view – in understanding and modeling cities, with important ideas about parameters and minimal models. We also develop a critical discussion of urban economics, and in particular of problems generated by the equilibrium assumption, and the existence of interactions. We will also discuss the important problem related to the choice of utility in these models. We end this chapter with a discussion about data, their sources and scales, and by discussing the necessity of interdisciplinarity in urban studies, including its current limitations that need to be overcome.

Chapters 3 to 8 are devoted to more specific subjects. In Chapters 3 and 4 we discuss the spatial structure of cities and transportation infrastructures, naturally opening the way to Chapters 5 and 6, which are devoted to mobility. Chapter 7 concerns the complex subject of the socioeconomic properties of cites, and in

Chapter 8 we advance a discussion of systems of cities and their hierarchical structure.

More precisely, in Chapter 3 we discuss the spatial organization of cities, and we will insist on their polycentric structure. We start with the problem of the optimal location of public facilities and retail stores, followed by a discussion on the empirical measures of polycentricity. We then turn to the theoretical side of polycentricity and review classical approaches such as the Fujita–Ogawa model and the "edge city" model proposed by Krugman. In particular, we show that these models are unable at this stage to predict correctly quantitative features such as the number of activity centers versus population. In order to understand this, we revisit the Fujita–Ogawa model and simplify it, obtaining results in agreement with empirical observations. This model, which describes in a simplified way where individuals are living and working, allows us to make predictions about mobility-related quantities such as CO_2 emissions or total commuting distance.

In Chapter 4, we discuss the structure and dynamics of transportation infrastructures. These objects evolve on very long time scales and only recently could we gather extensive data about their dynamics. We start this chapter by discussing the road and street network, which plays a central part in the morphodynamics of cities. In particular, its structure is intimately related to the distribution of activities and population. We then discuss the structure of subway networks and their time evolution. We also propose simple arguments that allow us to connect their network properties with economic aspects of their environment.

Mobility is the main subject of both Chapters 5 and 6. It is a crucial aspect of cities and is intimately related to the spatial organization of the city, its infrastructure, and the location of residences and activities. We have divided the various studies and discussions of this subject into two (related) chapters. In Chapter 5, we discuss patterns of mobility, described essentially by the origin–destination matrix, while in Chapter 6, we focus on mobility from the point of view of multimodality, described here in the framework of multilayer networks. More precisely, in Chapter 5 we discuss empirical results about the origin–destination matrix, obtained with the help of new data sources. We discuss how we can extract useful mesoscopic information from large, detailed datasets. In a second part, we will present the current theoretical understanding about mobility, starting from old approaches such as the gravity model to new theoretical discussions based on simple stochastic processes. In Chapter 6, we present another aspect of mobility, namely multimodality: there is, in fact, a growing trend in large cities to use different transportation modes. These different transportation networks are coupled to each other, and the recently developed framework of multilayer networks allows a new perspective on these systems. Within this framework, we discuss the effect of the coupling between the road and subway networks. We will also consider an information perspective on these networks and show that it will be increasingly difficult to find our way around large cities.

In Chapter 7, we discuss socioeconomic features of cities. This provides an opportunity to introduce classical models of urban economics such as the Alonso–Muth–Mills (AMM) model, the archetype of this type of approach. We believe that it is important to know and understand these models in order to test them empirically and also to be able to improve and to elaborate on them. Since much of this literature has been written by and for economists, we believe that a simple explanation of the main results for non-economists ought to be helpful and valuable for anyone interested in cities and their economics. In particular, we will go through all the derivations in these models in simple language accessible to scientists not trained in economics. These models, despite their many issues, constitute good starting points for other approaches that could be more elaborate, at least from the point of view of ingredients or mechanisms. In addition to the classical AMM model, we also discuss the Beckmann model that proposes a framework for understanding the interplay between the social network structure and the spatial organization of cities, a very real problem, thanks to the availability of social data. We also recount a recent discussion of the Schelling model for segregation and the problem of individual versus global optimization. We end this chapter with a discussion on how various macroscopic parameters describing specific features of cities vary with population. This allows us to introduce scaling properties for these systems, and we review here the main results and discussions of this topic.

In Chapter 8, we discuss models of systems of cities. The main purpose of these models is to explain the hierarchy of cities. This hierarchy is observed in the distribution of population as reflected by the Zipf law. We discuss in detail the mainstream models by Gibrat and Gabaix and also the model of diffusion with noise. This model – also called stochastic diffusion – is another regularization of the Gibrat model and likely represents a promising direction of research for understanding the coupling between cities. It is also a nice example of how modern approaches from statistical physics can be helpful for understanding complex systems such as cities. The hierarchy of cities is also reflected in their spatial organization, as discussed almost a century ago by Christaller in his central place theory. We present Okabe and Sadahiro's review of this famous approach, showing that fluctuations in a null random model could actually explain most of the observations made by Christaller. This is an interesting example as it makes the connection between a classical approach in regional science and properties of disordered systems. In particular, it highlights the importance of considering null models when discussing empirical data.

A final chapter is devoted to the outlook for the field, with discussions on testing qualitative ideas about cities and their future, and concluding thoughts about the possibility of a science of cities.

Of course these chapters cannot cover the whole variety of new studies about cities, and represent a small share of all city-related topics. Important studies in geography and economics are certainly missing, but what we have attempted here is to put together new approaches combining new tools and data in order to understand quantitatively as many aspects of cities as possible. This book thus inevitably covers a relatively subjective selection of topics. Owing to these personal biases, space limitations, and the sheer volume of current literature, important topics have been omitted. In this respect, I apologize in advance to those colleagues who feel that their work is not well represented here. Incomplete and imperfect as it is, I hope however that this book will be helpful to scientists interested in quantitative approaches to urban systems, a fascinating subject.

Acknowledgments

I joined the CEA in 1992 but moved to the Institut de Physique Théorique only in 2009; I warmly thank Yves Caristan who made this transfer possible. A believer in the value of interdisciplinary studies, he provided me with unstinting support. Henri Orland was at that time the director of the IPhT and I thank him for accepting me at the Institute. It is very probable that without him, my career evolution would have taken a very different turn and would probably not be as much fun as it is now.

I also thank Gabriele Fioni and Vincent Berger, former and current directors of the Direction des Sciences de la Matière (now Direction de la Recherche Fondamentale) and Michel Bauer, current director of the Institut de Physique Théorique, for creating such a great environment and providing the freedom necessary to tackle interdisciplinary studies. Likewise, I am grateful to the Centre d'Analyse et de Mathématique Sociales at the Ecole des Hautes Etudes en Sciences Sociales who hosted me regularly and provided a great multidisciplinary environment. In particular, I warmly thank its former and current directors, Henri Berestycki and Jean-Pierre Nadal, for their generous support and for some challenging discussions.

During my time as a student at the Ecole Normale Supérieure, Jean-Philippe Bouchaud was a constant source of inspiration for me. I still recall early talks about broad distributions and his lectures on statistical physics that always brought me deep and potent insights into disordered systems, the importance of fluctuations, and subtleties in stochastic processes. Another crucial tipping point in my career was my postdoctoral period in Gene Stanley's Center for Polymer Studies at Boston University. I learned many things there, and thanks to Gene's open mind and interest in new subjects, I came to understand the value of interdisciplinarity and the possibilites of exploring new subjects and new ideas and of trying what has not yet been done – a crucial ingredient in Science. This experience also taught me how to deal with empirical data, as well as the joys of extracting regular patterns in large amounts of data. Thanks to Shlomo Havlin I learned how to connect tools from statistical physics to other subjects. For all these reasons, I express my gratitude to Gene and Shlomo for having shown me the way.

Concerning cities, this book gives me the opportunity to thank Michael Batty for his continuous help and advice these last years. Through many discussions with Michael, I have benefited from his broad knowledge of urban systems, and I owe him a great deal for what I learned about cities. I also had the chance to meet and converse with Denise Pumain. I thank her for her supportiveness and her accurate insights into complex systems and cities, and for helping me to understand many issues.

I also thank all my collaborators and colleagues, together with my postdocs and PhD students with whom I worked on various subjects relating to cities. Constant interaction with other points of view is a vitally important ingredient in research, helping to shape one's ideas. Another important aspect, especially when working on cities, is interdisciplinarity. This brought me to meet many scientists from whom I learned much about completely different aspects of the field, ranging from applied mathematics and probability to economics, geography, and history. For all these discussions and interactions I warmly thank: E. Akkermans, E. Arcaute, A. Arenas, A. Barrat, A. Bazzani, I. Benenson, L. Benguigui, L.M.A. Bettencourt, G. Bianconi, A. Blanchet, P. Bordin, J. Bouttier, A. Brès, A. Bretagnolle, M. Breuillé, O. Cantu, G. Carra, A. Chessa, V. Colizza, Y. Crozet, M. De Nadai, S. Derrible, A. Diaz-Guilera, S. Dobson, A. Flammini, S. Fortunato, M. Fosgerau, E. Frias-Martinez, R. Gallotti, L. Gauvin, J. Gleeson, C. Godrèche, M. Gonzalez, M. Gribaudi, E. Guitter, A. Hernando, R. Herranz, H. Herrmann, P. Jensen, H. Jeong, K. Kaski, M. Kivela, J.P. Kropp, R. Lambiotte, V. Latora, J. Le Gallo, R. Le Goix, F. Le Nechet, M. Lenormand, T. Louail, R. Louf, J.-M. Luck, K. Mallick, C. Mascolo, C. Monthus, Y. Moreno, R. Morris, A.E. Motter, I. Mulalic, V. Nicosia, A. Noulas, J. Perret, M. Picornell, S. Porta, M.A. Porter, J. Portugali, D. Quercia, J.J. Ramasco, S. Rambaldi, D. Ribsky, C. Roth, C. Rozenblat, A. Ruas, M. San Miguel, F. Santambroggio, J. Saramaki, S. Shai, Y. Shibata, C. Sire, E. Strano, G. Theraulaz, P. Vertes, A. Vespignani, M.P. Viana, A. Vignes, and H. Youn.

I also acknowledge the generous funding support at various stages of this interdisciplinary research by the European Commission (Plexmath and EUNOIA), the Centre National de Recherche Scientifique (PEPS-MOMIS program), the Société du Grand Paris (SGP) and the city of Paris through its program "Paris 2030". I would like to thank the staff at Cambridge University Press for their excellent support and responsiveness. In particular, I thank Nick Gibbons for his constant support and for many discussions about this project. I also thank Roisin Munnelly for guiding me through the whole book-production process, and Tom Moss Gamblin, my copy editor, for his crucial assistance with the pitfalls of the English language.

For everything, I thank my loving family, Esther, Rebecca, and Catherine.

1
Urban systems

In this first chapter, we propose a rough, synthetic view of cities, by retaining what we believe to be some salient features from a quantitative point of view. There are many books and reviews, giving countless details and figures about cities (in particular, the reader can consult the updated version of reports produced by the UN – see for example "World urbanization propects, the 2014 revision"), and instead of offering a long list of various properties of cities (which can be found in different books and reliable sources such as the Census Bureau for the US, the UN, the OECD, or the World Bank), we focus here on a small set of key figures and discuss important scales manifesting in cities.

Cities are complex objects with many different temporal and spatial scales, related to a large number of processes. While a small set of numbers is certainly not enough to describe the full complexity of cities, such numbers can nevertheless allow for quantitative studies and for a large-scale characterization of urban systems. There is much variety among cities in terms of morphology, population, density distribution, and also functions, yet despite these differences, we observe statistical regularities for some observables. Indeed, we can expect that large systems composed of a large number of constituents lead to collective behaviors characterized by statistical regularities. Another reason for this "universality" is the existence of fundamental processes common to all cities: spatial organization of activities and residences, mobility of individuals, and so on. One of the most challenging problems of a science of cities is then to identify the minimal set of mechanisms that describe the evolution of cities.

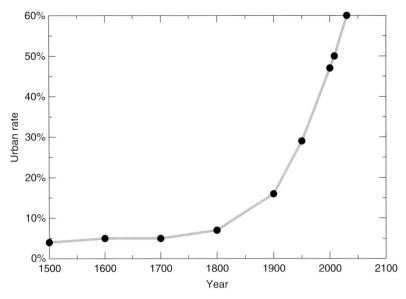

Figure 1.1. Evolution of the global urban rate (data from the HYDE history database http://themasites.pbl.nl/tridion/en/themasites/hyde).

1.1 A science of cities

1.1.1 The nature of the problem

A central issue in understanding urbanization is the large number of entangled, time-varying processes that generate cities. Many disciplines such as quantitative geography or urban economics have addressed some aspects of this problem and produced either very abstract models, or, at the other extreme, simulations with very large numbers of parameters designed for specific locales.

A proposal for a (new) science of cities that has recently emerged is based on an interdisciplinary strategy using ideas and tools from statistical physics of disordered systems, quantitative geography, and spatial economics. Key to this strategy is the extraction from data of universal facts that go beyond specific historical or geographical aspects of cities. These results will then be used to construct models with a minimal number of microscale ingredients leading to emergent collective behavior consistent with data.

This will be a landmark step for the construction of a science of cities and will allow for a long-delayed comparison of theory with empirical data. The results will impact many areas concerned with urban systems and will have practical applications for the planning of future cities. In particular, the primary goal is to understand the object "city" and to identify the dominant forces that shape its formation and evolution. Beneath these approaches lies the idea that the concept of

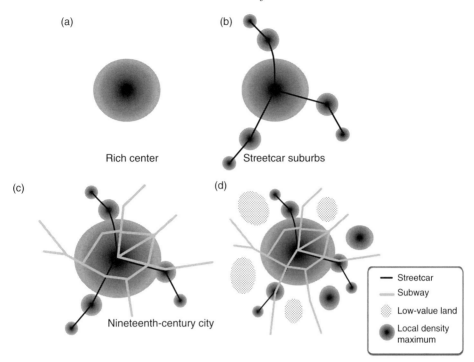

Figure 1.2. Illustration of the different phases for American cities (see for example Anas et al. (1998)). (a) Before the advent of streetcars and subways, businesses and manufactures concentrated in the center of the city, leading to a monocentric city with richer people living in the center. (b,c) Streetcars and subways allowed commuters to live further away from the CBD and around stations. (d) Cars increased mobility and former, cheap non-accessible zones rapidly became residential areas.

a city has some reality, albeit abstract, and that what we observe in the real world are simply specific instances of this entity on particular substrates, geographies, and histories.

The goal of a science of cities will be reached when, considering a specific case, we can basically say what will happen and which ingredients it is necessary to introduce in a model in order to get more detailed information and predictions.

First of all, we have to agree on the meaning of the word "understanding." For example, in the remarkable paper by Anas et al. (1998), there is a clear description of the time evolution of typical American cities (see Fig. 1.2): the high cost of communication led to concentration of industry and before 1850, personal transport was mainly by foot, resulting in a typical monocentric structure around the central business district (CBD) with income decreasing with distance from this center. The advent of electric streetcars led to the so-called "streetcar surburbs" structure where commuters live around stations that are radially dispersed from the CBD, a pattern that was later reinforced by the construction of subways. The automobile, which rapidly became available to the middle class, increased the mobility of individuals, and the low-value interstitial areas between streetcar (or subway) lines quickly became important residential areas. There is therefore a strong correlation between the transportation mode available to a large fraction of the population and the urban structure. The fact that different transportation systems appeared at different times, together with the durability of infrastructures, can then explain several patterns: older cities in the United States, on the East coast, have streets and buildings dating back to the era of their development as harbors, while more recent cities that developed during the automobile era have a spatial structure essentially determined by their highway systems. European cities have usually developed around medieval towns, but in contrast with US cities, the centers are usually still a mixture of residential and business areas, probably resulting from the large number of cultural amenities there.

This discussion shows that we "understand" some of the mechanisms leading to the formation of a city and governing its evolution. Here, understanding means that we can construct a consistent story, based on a few ingredients, that explains a selection of facts observed in reality. While this is somewhat satisfying, it is not enough for constructing a science of cities: indeed, we would like to assess quantitatively the impact of various factors, which means that we want to write mathematical relations between different quantities. The main point here is to identify the most important parameters, not only to understand the past, but also to be able to construct a model that indicates with reasonable confidence the future evolution of a city and allows the impact of various policies to be tested.

1.1.2 What is a city? Origins and definitions

From a very general point of view, cities exist in order to connect people. This primacy of interactions has now been acknowledged by many scientists, ranging from geographers (Batty 2013) to urban economists (see for example Thisse 2014). As pointed out by Brueckner et al. (2011), the explanation for the existence of cities varies a great deal from one field to another, depending on inclination. One goal of a science of cities is then to encompass these inclinations and to identify the driving mechanisms responsible for their formation and evolution.

It is still true to say that individuals interact face-to-face, and it is reasonable to follow urban economics that identify economies of scale and agglomeration effects as crucial factors in the formation of cities (see Duranton and Puga, 2004). Indeed, the total number of possible interactions between individuals grows very quickly as $P(P-1)/2$ (and probably saturates toward a finite value that corresponds to the Dunbar number of stable relationships; see Dunbar, 1992 and the recent validation via Twitter data, Gonçalves et al., 2011). Between two distinct groups of size P_1 and P_2 the number of possible interactions is $P_1 P_2$ and can thus also be very large. For a region of linear dimension R and uniform density ρ, the maximum number of possible direct interactions thus scales as $\rho^2 R^4$, showing the very strong effect of spatial localization and density. The desire to be at the same location, however, naturally induces competition and the appearance of a real-estate market.

Transport technology is the medium making these cities possible and is thus of crucial importance in the formation and evolution of cities. In addition to mobility infrastructure, we can list all the main elements that govern the formation and the evolution of cities: A network of interactions between individuals, flows of goods between firms, and infrastructure networks, making these interactions and flows possible. In addition, a city is never isolated but rather is part of a system of cities, at the national level, and now also at a global level (Bretagnolle et al. 2009).

There are obviously many other ingredients, such as policies, governance, and so on, and the numerousness of these elements is part of the problem in constructing a science of cities. Understanding how these different elements combine with each other and govern the evolution of urban systems will inevitably necessitate filtering some of these ingredients while keeping the most relevant of them. This huge variety of problems relating to cities concerns many different disciplines ranging from geography to applied mathematics, to transport, to urban economics, to qualitative social sciences. It is thus impossible to discuss all aspects of cities at once, casting some doubts on the possibility of a "unified theory" of cities able to describe all aspects of their evolution.

Even the apparently simplest question of how to define a city rapidly leads to significant problems. There are indeed various definitions of cities depending on era or country. The administrative definition, although straightforward, is outdated and does not capture urban sprawl. In order to remedy this, various agencies have proposed other defintions such as urban areas, MSAs (metropolitan statistical areas, in the US, FUAs (functional urban areas, as per the OECD), and LUZs (larger urban zones, in Europe), which basically define urban centers and connect them to other areas for which the commuter fraction is larger than a given threshold. These zones are defined such that they incorporate functional areas, but unfortunately are not necessarily consistent with each other. In addition, each of these definitions relies on a (small) set of parameters and it would be better to find a definition that can be used across time and also across space in order to compare different cities in different countries.

Another form of definition relies on non-ambiguous objects such as built-up areas and on the notion of contiguity. For example, Rozenfeld et al. (2008, 2011) introduced a bottom-up method for constructing cities by clustering areas from high-resolution data. More precisely, the City Clustering Algorithm (CCA) is

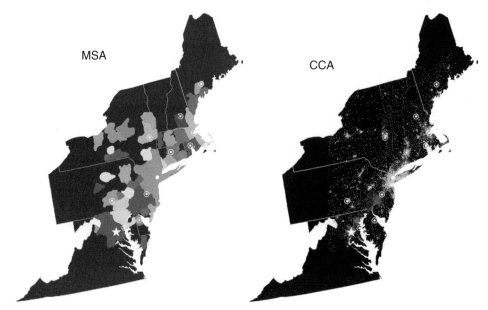

Figure 1.3. MSA for the northeastern US and the clusters obtained by the CCA algorithm with $\ell = 5$ km (the white circles denote the location of state capitals). Figures from Rozenfeld et al. (2011).

defined as follows: we first locate a populated area and then grow the cluster by adding all populated sites within a distance smaller than some level ℓ and with a population density larger than a threshold ρ^*. The cluster stops growing when no site at a distance less than ℓ and with density larger than ρ^* can be added. Rozenfeld et al. (2011) chose ρ^* and took ℓ in the range [0.2, 4] (in km). It is interesting to note that this procedure leads to clusters that are not too different from actual MSAs (see Fig. 1.3 for an example).

Elaborating on these various ideas of percolation and functional areas, Arcaute et al. (2013) proposed a variant of the CCA algorithm to define cities. They use population density as the main parameter and start from a given unit of agglomeration ("wards" in the case of the UK). For a given unit, they cluster all adjacent units with density larger than a threshold ρ^* (see Fig. 1.4). Interestingly, there is a kind of "percolation transition" for a value of $\rho^* \approx 14$ persons per hectare above which cities merge together (in the instance, Liverpool and Manchester) to produce a very large cluster which contains the majority of the population (> 70%) and represents the majority of the total area (> 50%). For this density, the distribution of population follows a power law with exponent close to 2 as expected from Zipf's law and the clusters for this density threshold are therefore a good candidate for defining cities.

Figure 1.4. Cities obtained by the clustering algorithm for different density cut-offs. From top left to bottom right: $\rho^* = 40\,prs/ha$, $\rho^* = 24\,prs/ha$, $\rho^* = 10\,prs/ha$, and $\rho^* = 2\,prs/ha$. Figure from Arcaute et al. (2013).

In a second step, Arcaute et al. (2013) incorporate the notion of functionality through the number of commuters. Once they have detemined the clusters by density, they consider them as destinations of commuter flows (as long as their population is larger than a threshold N, in order to select important commuting hubs only) and add areas that are the origins of these flows. For each given ward, they then compute the fraction of individuals that commute to each of the destination clusters and the ward is added to the cluster that receives the largest flow if the flow is above a threshold τ_0. This procedure allows construction of cities with two parameters (N, τ_0) and testing, for example, the robustness of scaling exponents (see Section 7.4).

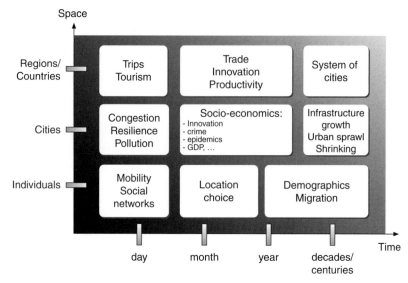

Figure 1.5. Schematic representation of processes occurring in urban systems according to their temporal and spatial scales.

1.2 Spatial and temporal scales

There are many different spatial and temporal scales in a city, each related to various processes occurring in these systems. It is not only the diversity of these scales that make the modeling of cities difficult, but also the large variety of agents and their impacts on the life and evolution of urban systems. We represent in Fig. 1.5 the most important processes and the order of magnitude of the spatial and temporal scales involved. We see that we have a wide spectrum of scales that are mixed together. This almost continuous spectrum makes it difficult to consider cities as being in equilibrium (see Chapter 2) but also to view these processes as decoupled from each other. This is a very important problem in city modeling and necessitates a careful discussion of spatial and temporal scales, acknowledging the possibility that various processes are interfering with each other.

We will now describe in more detail some typical scales and variations of important quantities for cities, such as population, area, and (population) density.

1.2.1 Population

Population is a critically important parameter for cities. Indeed, it is often assumed to be the explanatory variable, neglecting endogeneity issues. In many cases, however, knowing the population of a city reveals much about it (Pumain and Moriconi-Ebrard 1997) and seems to be a good starting point for a theoretical framework to understand how cities evolve as their populations grow. Eventually,

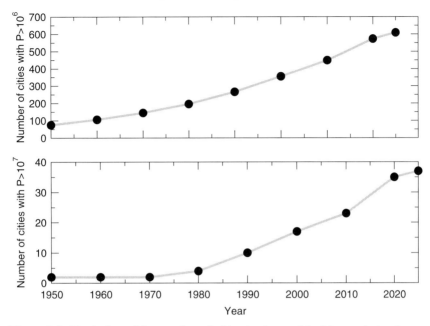

Figure 1.6. Evolution of the number of cities in the world with population larger than 1 million (top) and the number of megacities (with $P > 10^7$ inhabitants; bottom). Data from the UN.

a theory that endogenously describes the evolution of cities and their populations would mark a fantastic advance in our understanding of these systems.

We observe a large variety of cities around the world, with very different populations and growth rates. In this section we present a quick overview of the main facts regarding this quantity.

Size of cities

There is a wide variety of city sizes, from small towns with 10^4 inhabitants to megacities with populations greater than 10 million. Large cities and in particular megacities represent almost 10% of world urban population, and their number is growing (Fig. 1.6). We expect approximately 40 megacities in the world by 2020.

Concerning the structure and evolution of cities, we need to separate the developed countries from the developing ones. According to the UN Statistics division (see for example the Demographic Yearbook and World Urbanization Prospects for 2005 and 2011), for developed countries there is a decline in native-born population growth and it is estimated that about 1/3 of urban growth will come from migrations. The urban population – as shown in Fig. 1.7 – is essentially concentrated in small cities (in the range $[10^5, 10^6]$ inhabitants) with 30.5% living in cities with more than 1 million inhabitants. This has to

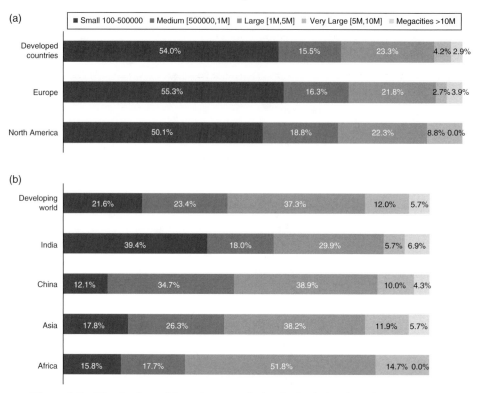

Figure 1.7. Distribution of the urban population in (top) developed countries and (bottom) developing countries. Data from the UN Statistics Division.

be contrasted with developing countries where there is still a strong pace of urbanization with an average of about 2% per annum (2015). Also, in these countries, urban dwellers live preferentially in large cities, with 54% of the urban population in cities with more than 1 million inhabitants.

Growth rates

Urban growth has different sources such as migration or birth. In developed countries, international migration will become the main source, while for developing countries it is the high birth rate that increases the city population. In China, however, the main source of growth is the migration from rural to urban areas. The growth rate of cities depends on their size; in Table 1.1 we show the results obtained by Bretagnolle et al. (2009). More precisely, if we distinguish developed from developing countries, we see that growth rates are distributed very differently. For developed countries, in a total of over 1,287 cities (with size larger than 100,000 inhabitants), 39.9% are declining, while 34.9% of the 1,398 developing-country cities (with $P > 100,000$) are experiencing rapid growth (see Fig. 1.8). For example, China in 2025 will have approximately 140 cities with more

1.2 Spatial and temporal scales

Table 1.1. *Growth rate (%) of cities according to their sizes for European, Indian cities, and US (Russian not being considered here). Data from Bretagnolle et al. (2009).*

Population	Europe (since 1850)	India (since 1901)	USA (since 1940)
> 100,000	1.38	2.45	1.79
[50,000 − 100,000]	1.13	1.95	0.92
< 50,000	0.99	1.71	–

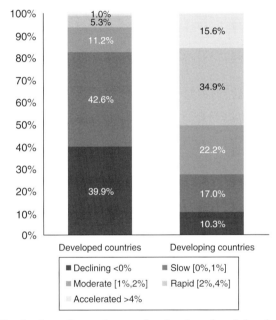

Figure 1.8. Distribution of growth rates for developed and developing countries (data from the UN).

than 1 million inhabitants, while in India, urban growth will be greatest in medium-sized cities and in the suburbs of large cities.

It is however important to note here that even if growth rates are smaller for large cities, in terms of numbers of individuals large cities will still have to absorb huge numbers. These simple observations show that the dynamics of cities display various regimes, an important fact that models will have to explain.

Population distribution and Zipf plot

A classical observation about cities concerns the distribution of their populations. Far from being peaked – as a theory of optimal size would predict – the distribution we observe is broad with fat tails. Known as Zipf's law (see various examples in Zipf, 1949), the rank-ordered plot (the "Zipf" plot) of populations displays a "universal" behavior of the form

$$P_r \sim r^{-\nu} \tag{1.1}$$

where ν is an exponent whose value is often around 1. In this expression, we have sorted a collection of population P_i ($i = 1, \ldots, N$), and renamed them such that $P_1 > P_2 > \cdots > P_N$. The index thus corresponds to the rank r, and the Zipf plot is simply P_r versus r. This result is easily related to the probability distribution of the quantity P. Indeed, the rank is essentially $r(y) = NF(y)$, where $F(y) = \text{Prob}(P > y)$ (one can check that for the maximum $F(P_{max}) \sim 1/N$, leading to $r = 1$), and Zipf's law implies that the population P is distributed according to a power law of the form

$$\rho(P) \sim \frac{1}{P^{1+\mu}} \tag{1.2}$$

with the cumulative function behaving as $F(y) \sim y^{-\mu}$, where

$$\frac{1}{\mu} = \nu. \tag{1.3}$$

A value $\nu \approx 1$ leads then to an inverse power law for the population distribution with exponent $1 + \mu \approx 2$ (for the value of the exponent see for example Soo 2005; Rozenfeld et al. 2008; Malevergne et al. 2011). For US and European cities, for example, we observe different exponents, as shown in Fig. 1.9. In fact, the exponent μ is not universal and displays important fluctuations from a country to another as shown on its distribution in Fig. 1.10.

As we discuss in detail in Chapter 8, the Zipf law has important consequences on the number of cities and the largest city in a given country (Pumain and Moriconi-Ebrard 1997). In particular, it implies that both the number of cities and the size of the largest city are essentially proportional to the total urban population. An important theoretical challenge is to understand the origin of the Zipf law and in Chapter 8, we discuss various models that were proposed in order to explain its existence.

1.2.2 Area, density, and volume of cities

Area

A city may be defined, first, by its area A, whose definition unfortunately is not unique across time and across countries. As discussed above, a reasonable definition relies on contiguity between built-up areas (Rozenfeld et al. 2008; Arcaute et al. 2013) and the resulting giant cluster can be considered as being the city. The length $L = \sqrt{A}$ is an important distance to compare to various other

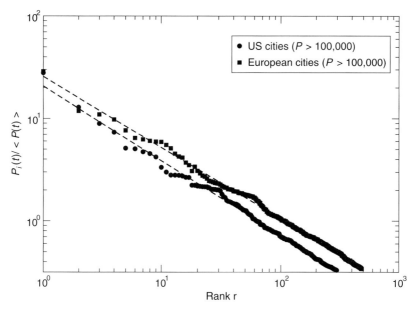

Figure 1.9. Zipf plot (population versus rank) for US and European cities with population larger than 100,000 (here the population is normalized by the average allowing to compare different times). For the US cities, the population is for 2014 and estimated by the US Census Bureau. For the European cities, the data are obtained from various sources and compiled at the City Population website (www.citypopulation.de). The dashed lines are power-law fits giving exponents for the US and Europe: $\nu_{US} \approx 0.73$ ($r^2 > 0.998$), $\nu_{EU} \approx 0.70$ ($r^2 > 0.998$).

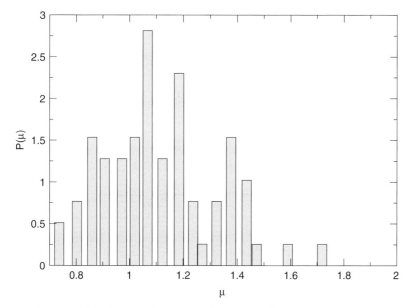

Figure 1.10. Distribution of the exponent μ (computed by ordinary least-square estimate) for 72 countries in the world (data from Soo 2005). The average is $\overline{\mu} = 1.11$ and the dispersion 0.20.

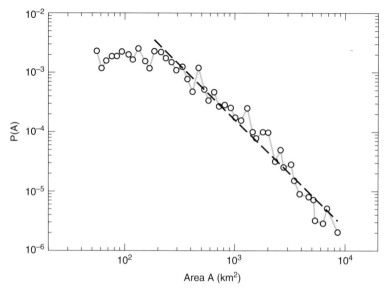

Figure 1.11. Urban area distribution for cities worldwide with population larger than 1 million. The dashed line is a power-law fit with exponent $\tau \approx 1.83$ ($r^2 = 0.98$) (data from *Demographia, World Urban Areas*, 12th Annual edition (2016)).

scales, typical orders of magnitude ranging from $L \sim 1$ km for small cities to very large cities with an extension of order $L \sim 100$ km (e.g. Los Angeles). For cities with population greater than 1 million the area varies over the wide range of $[41, 11,642]$ km^2. We thus see that cities differ enormously in area, and in Fig. 1.11, we show the (urban) area distribution for the 494 cities worldwide with population larger than 1 million. This shows clearly that area is a very broadly distributed variable and a power-law fit gives a behavior for the tail of the form $P(A) \sim A^{-\tau}$ with $\tau \approx 1.83$. This behavior is also valid at a smaller scale: using the CCA algorithm, Rozenfeld et al. (2008) found an exponent of order 2 for the United States and the United Kingdom (see Fig. 1.12).

Population density

The average population density defined by

$$\rho = \frac{P}{A}, \qquad (1.4)$$

where P is the population of the city and A its area, is an important parameter for distinguishing different types of cities (see below). The inverse density $a = 1/\rho$ is the average area per capita and represents the order of magnitude of the area available to each inhabitant (although this is a very rough measure, as we should take into account tall buildings and only the residence area, which can be much smaller than the total area after industries and infrastructures are excluded). The

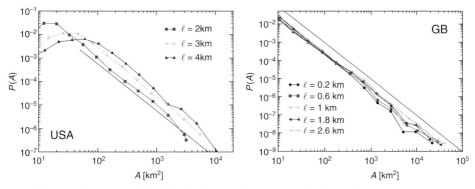

Figure 1.12. Probability distributions of the areas of cities found by the CCA algorithm with different values of ℓ for (a) the United States and (b) the United Kingdom. The solid straight line corresponds to a power law with exponent -2. The figures are taken from Rozenfeld et al. (2008).

order of magnitude of population densities varies from one region of the world to another. For the 494 cities with population larger than 1 million inhabitants, we plot the fraction of world regions for a sliding window of 10 cities; the result is shown in Fig. 1.13. We thus see a clear distinction between different types of cities according to their geographical location:

- Asian cities are the densest, with $\rho \in [20,000; 40,000]$ ha/km^2.
- North American cities have a low density up to 2000–3000 ha/km^2 (which is consistent with the fact that the main transportation mode is the car).
- The intermediate case corresponds to European cities with values in the range $\rho \in [1,000; 10,000]$ ha/km^2.

How the density varies with population is however not clear. In order to understand this relationship, we can plot the area versus the population for different countries. We see that we can fit the data with a linear function of the form

$$A = aP \tag{1.5}$$

where a is the inverse density. We could also fit the data by a nonlinear function of the form

$$A = aP^\tau \tag{1.6}$$

where τ is an exponent usually slightly smaller than 1. In the first case (Eq. (1.5)) the density is constant, $\rho = P/A = 1/a$, while in the nonlinear case (Eq. (1.6)) the density increases with population $\rho \sim P^{1-\tau}$. We observe that countries actually display very different behaviors. In the case where we mix all countries of the world by taking urban areas with population larger than 1 million (Fig. 1.14(a)), the nonlinear fit predicts an exponent slightly less than one

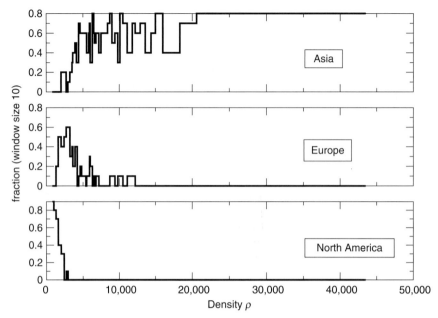

Figure 1.13. Fraction of world regions vs. density. We ranked in decreasing order world cities (with population greater than 1 million) according to their density and computed the fraction of each region (Asia, Europe, or North America) for a sliding window of size 10. We see that the majority of the most densely populated cities are in Asia, intermediate densities are found in Europe (the Russian Federation is considered to be in Europe here), and low-density cities are typical for North America.

($\beta \approx 0.95$, $r^2 = 0.66$), and with this dataset it does not seem possible to decide between a constant density or a slightly increasing function of population (with exponent of order 0.05). At a smaller scale, in the case of the United States for example, the nonlinear behavior is also indistinguishable from the linear one. The linear fit predicts an average area of order 660 m^2, which is large, while the nonlinear fit gives an exponent $\beta \approx 0.85$. While the linear fit predicts a constant density of order $1,500$ ha/km^2, which is correct for many cities, we actually see a deviation for large cities, with density much larger for cities proper (and not for urban areas that mix heterogeneous zones). For example, we have densities of order $10,000$ ha/km^2 for New York City, or $6,600$ ha/km^2 for San Francisco.

This behavior is however completely different for a country such as Japan (Fig. 1.15), where both linear and nonlinear fits are poor. The noise in this case permits no firm conclusion and new theoretical insights are certainly needed here in order to understand these empirical observations.

1.2 Spatial and temporal scales

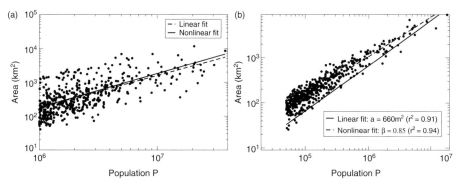

Figure 1.14. Area versus population for different regions in the world. (a) Area of the 494 world cities with population larger than 1 million inhabitants. Both the linear and nonlinear fit are acceptable (data from various sources compiled by *Demographia, World Urban Areas*, 12th annual edition (2016)). (b) US urban areas with population over 50,000 (data from the 2010 Census population).

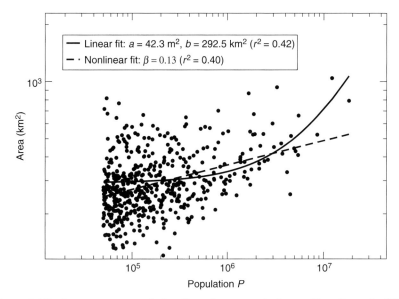

Figure 1.15. Area versus population for urban areas in Japan (data from the UN, Demographic Yearbook 2012). The linear fit is of the form $A = aP + b$ and the nonlinear fit of the form Eq. (1.6).

We end this section by noting that a more accurate quantity is the local density defined over a more restricted area:

$$\rho(x) = \frac{P(x)}{A(x)} \tag{1.7}$$

where $A(x)$ is an area surrounding point x and $P(x)$ the corresponding population. The variations in $\rho(x)$ represent an important aspect of cities and were analyzed in many studies (see for example Bertaud and Malpezzi 2003). Even if the average density is the same, we can observe different organizations of cities with one density peak, or in contrast a more complex organization with a polycentric structure. We will return to this aspect in Chapter 3, on the spatial organization of cities.

Volume of cities

A more accurate image of cities emerges when we take into account their three-dimensional aspect. Indeed, by considering the average density we implicitly assume that individuals are distributed over a two-dimensional surface. A building is however a volume and can be roughly characterized by three parameters: its footprint area S with perimeter p, and its height H. The volume of the buiding is then essentially $V = SH$ (which is correct for a parallelepiped shape) and we can naïvely expect the following scaling:

$$H \sim p \tag{1.8}$$

$$S \sim p^2 \tag{1.9}$$

$$V \sim p^3, \tag{1.10}$$

Batty et al. (2008) studied in detail how these different quantities are distributed for cities around the world and in particular for Greater London. For a quantity X (which can be the surface, perimeter, or volume), we define the exponent κ_X that characterizes the decay of its probability distribution as:

$$\rho(x) \sim \frac{1}{x^{1+\kappa_X}}. \tag{1.11}$$

For Greater London, the results are somewhat surprising. First of all, depending on the building use (residential, office, retail, etc.) the exponents κ_X display some variations. In all cases, however, these exponents are greater than 1 and only for the surface and volume is the exponent between 1 and 2. This means that these quantities display a very large variance, while the perimeter and the height of the building have a tail which decays quickly (and the distribution has a finite variance).

Beside their distribution, it is important to understand how these quantities vary with city size (measured by population). Recently, Schläpfer et al. (2015) obtained data on the heights of buildings in 12 US cities and observed that the average height $\overline{H}$ scales with the population as

$$\overline{H} \sim P^\beta \tag{1.12}$$

where β here is around 0.16. Schläpfer et al. (2015) proposed a scaling theory for this exponent and we give here a simple argument that predicts its value. Indeed, if we assume that the volume of apartment space per capita is constant (given by v_0), then we have

$$\frac{\overline{H}}{v_0} \frac{A}{P} \sim 1 \qquad (1.13)$$

For the United States, we know that the area of a city scales as $A \sim P^a$ (with $a < 1$ and $a \approx 0.85$, see above), and we obtain

$$\overline{H} \sim \frac{P}{A} \sim P^{1-a} \qquad (1.14)$$

which leads to $\overline{H} \sim P^{0.15}$ in agreement with the empirical observations of Schläpfer et al. (2015).

1.2.3 Time scales

There are many time scales at work in cities; one of the smallest corresponds to daily mobility. Since the work of Zahavi (1974); Marchetti (1994), it is often assumed that there is a time budget of about one hour for daily mobility – sometimes called Marchetti's constant. This "rational locator hypothesis" assumes that individuals maintain constant journey-to-work travel times by adjusting their home and workplace. However, recent measures showed (Levinson and Wu 2005) that commuting times increase in time at the intra-urban level and clearly depend on the metropolitan spatial structure. From a theoretical point of view, commuting time increases with city population, as an effect of congestion (see Section 3.4). Measures in US cities indeed showed that the time spent in traffic jams grows with city size (Louf and Barthelemy 2014a), giving another argument against the invariance of daily mobility time. The time budget evolves much more slowly than the spatial "horizon" of cities, however, and as technology and transport infrastructure improve, the horizon of the city evolves over unprecedented scales, while isochrone lines become strongly anisotropic and follow the transport infrastructures.

Larger time scales are observed for urban growth, with essentially two sources: birth and migration. The global average urban growth rate is 1.8% but there is a relatively large dispersion around this average, revealing large geographical disparities. As shown in Fig. 1.16, urban growth rates vary from negative values to large positive ones. A closer inspection reveals that negative and small values are observed primarily in European countries, while the large values are observed mainly in Africa, with growth rates of order 6% for Burkina Faso or Uganda for example.

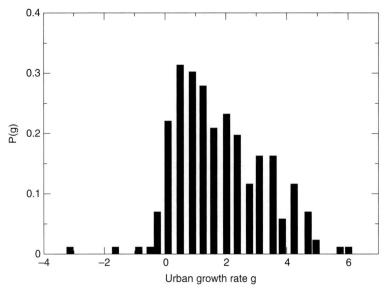

Figure 1.16. Distribution of the urban growth rate in different countries in the world (data from http://world.bymap.org/UrbanPopulationGrowthRates.html).

We note that a yearly growth rate in the order of 5% corresponds to a typical time scale of 20 years which governs the variation of population in cities. This order of magnitude is consistent with the value obtained by a correlation analysis (see Chapter 8 and Hernando et al. 2014) which gives a correlation time of order 25 years for the United States. Finally, there are also time scales associated with the evolution of infrastructures such as roads and subways (see Chapter 4). These scales are typically of the order of at least a year and describe a process that can last many decades.

1.3 Naïve scaling

In many theoretical approaches discussed in this book, the population is considered as being the main explanatory variable, and modeling essentially consists in understanding and predicting the evolution of various quantities with this variable (as discussed above, considering that taking population as the explanatory variable is a simple way to attack this problem). We present here simple arguments explaining how various quantities such as area A, total daily distance driven L_{tot}, or total lane miles L_N vary with population size. There are essentially two main ingredients when considering naïve scaling. The first is dimensional analysis (see for example Barenblatt 1996), which implies that in any equation, dimensions should be consistent as a necessary condition for its validity. This aspect, sometimes overlooked, is however essential and can actually lead to a better understanding of how different quantities are related to each other. The second ingredient is the use of simple "null" models to provide benchmarking. Empirical deviations with respect to a benchmark offer valuable information about relevant mechanisms. Although these predictions turn out in general not to be correct, naïve scalings are nevertheless useful as a first approach to the problem, enabling an understanding of how different quantities relate to one another.

1.3.1 Surface area

We would like to estimate the dependence of the area A of a city on its population P, a long-standing problem in the field (Makse et al. 1995). A first crude approach is to assume that cities evolve in such a way that their average population density $\rho = P/A$ remains constant. This assumption straightforwardly implies that the area scales linearly with population:

$$A \sim \lambda^2 P \tag{1.15}$$

where λ^2 is the average area occupied by each individual (the assumption of a constant density is equivalent to a constant average area per capita). We can probably expect this behavior to hold for relatively small cities, where there is a constant-rate increase in the built-up area as the population increases. However, this regime will certainly be limited, since the city cannot extend indefinitely, and we will then observe a behavior characterized by more slowly increasing area, for instance a power law with exponent less than one.

In Table 1.2, we compare naïve exponents with empirical measures for the United States. We see that in America, areas of cities do not scale linearly with population, and we observe a behavior consistent with a power law of the form $A \sim P^a$ with $a = 0.85$. This behavior implies that the average density $\rho = P/A$ behaves as

$$\rho \sim P^{1-a} \tag{1.16}$$

which confirms the common belief that the larger a city, the denser it is.

1.3.2 Total length of roads

The total length L_N of all the roads within a city is an important feature. It characterizes accessibility to a large extent, and can shed some light on the organization of the city. In general, this quantity is related to properties of spatial networks (Barthelemy 2011) and can be easily estimated given simple assumptions. Indeed, if we consider that the network formed by streets is such

Table 1.2. *Value of the exponent governing the behavior with the population P obtained by naïve arguments and the value obtained from empirical data for the US (data from the Census Bureau www.census.gov, urban mobility report http://mobility.tamu.edu/ums, and the Federal Highway Administration, https://www.fhwa.dot.gov).*

Quantity	Naïve exponent	Measured value
A	1	0.85 ± 0.011 ($r^2 = 0.93$)
$L_N/\sqrt{A}$	0.5	0.42 ± 0.02 ($r^2 = 0.83$)
L_N	1	0.86 ± 0.02 ($r^2 = 0.92$)
$L_{\text{tot}}/\sqrt{A}$	[0.5, 1]	0.60 ± 0.03 ($r^2 = 0.90$)
L_{tot}/P	0	0.03 ± 0.02 ($r^2 = 0.04$)

that each node (intersection) is connected to its closest neighbor, the typical length of a road segment is given by

$$\ell_R \sim \sqrt{\frac{A}{N}} \qquad (1.17)$$

where N is the total number of intersections. Previous studies of road networks in different regions and over extended time periods (Strano et al. 2012; Barthelemy et al. 2013) have shown that the number of intersections is proportional to population size (where the proportionality factor depends on the region considered). Therefore, the typical length of a road segment (between two intersections) varies with the population size P as

$$\ell_R \sim \sqrt{\frac{A}{P}} \qquad (1.18)$$

and the total length of the network $L_N \sim P\ell_R$ should then scale as

$$\frac{L_N}{\sqrt{A}} \sim \sqrt{P}. \qquad (1.19)$$

Using the naïve scaling for the dependence of A on population size given previously in Eq. (1.15), we obtain

$$L_N \sim P. \qquad (1.20)$$

We observe in Table 1.2 that this naïve scaling is not too far off, even if there seems to be a significative deviation from linear behavior for large cities.

1.3.3 Total daily commuting distance

The total commuting distance L_{tot} corresponds to the sum of commuting distances for all individuals in a given city and is usually described in person-kilometers. This quantity is crucial when individual cars dominate commuting: it is a signature of mobility, and is connected to the amount of gasoline consumed, CO_2 emitted, and time spent in traffic. The first constraint on this distance comes from individuals' limitations and behavior, and the simplest assumption is that the commuting distance per capita is constant. This corresponds to the case where individuals choose their home or their job at a distance around a typical value independent from the city size. Within this assumption, we simply have

$$\frac{L_{tot}}{P} \sim \text{const.} \quad (1.21)$$

An additional contraint on L_{tot} is given by the structure of the city (Samaniego and Moses 2008). Indeed, the individual commuting distance is also related to the total land area of the city and the location of activity centers. If we first assume that the city is monocentric, individuals are all commuting to the same center and the commuting distance is controlled by the typical size of the city of order $\sqrt{A}$,

$$\frac{L_{tot}^{mono}}{\sqrt{A}} \sim P. \quad (1.22)$$

On the other hand, if we assume that the city is completely decentralized, the typical commuting distance is on the order of the nearest-neighbor distance $\sqrt{A}/\sqrt{P}$, and we obtain

$$\frac{L_{tot}^{dec}}{\sqrt{A}} \sim \sqrt{P}. \quad (1.23)$$

Even if this is reasonable for the average value, there could however be non-negligible fluctuations. Indeed, as we shall see in Section 5.2, the commuting distance distribution $P(r)$ is actually broadly distributed and decays as $1/r^3$. The existence of large fluctuations then certainly precludes the simple argument's being perfectly correct.

If we test these naïve scalings on empirical data we observe that L_{tot}/P can be considered reasonably independent of P (with a value of approximately 23 miles for the US), in agreement with the individual constraint assumption (Eq. (1.21)). This finding is also in agreement with the results drawn from census data in Germany by Wilkerson et al. (2013).

The scalings of $L_{tot}/\sqrt{A}$ given in the extreme cases of a monocentric city structure and a totally decentralized city structure disagree with the value obtained from data (see Table 1.2), which suggests that most American cities have an intermediate "polycentric" structure characterized by a variable number of centers. Indeed, if we assume a city with k centers, characterized by their own basins of attraction with the same area A_1 and population P/k, we have

$$A = kA_1. \tag{1.24}$$

The total commuting distance is then given by

$$L_{\text{tot}} \sim k \times \frac{P}{k}\sqrt{A_1} \sim P\sqrt{\frac{A}{k}} \tag{1.25}$$

and if we assume that the number of centers varies with population as $k \sim P^\alpha$, we obtain

$$\frac{L_{\text{tot}}}{\sqrt{A}} \sim P^{1-\alpha/2}. \tag{1.26}$$

The exponent α is *a priori* less than one and we thus have an exponent value between 0.5 and 1, in agreement with measures obtained in the United States (Table 1.2 and Samaniego and Moses 2008). We note that Equations (1.16), (1.21) and (1.26) are indeed consistent and would lead to a value of $\alpha \approx 0.8$ for the US. This result on polycentric structure is obviously very important and possible explanations are presented in detail in Chapter 3.

We see from this discussion that important information about the structure of cities and mobility patterns can be encoded in the nontrivial values of exponents characterizing the evolution of macroscopic quantities such as area, total commuting distance, and so on. In this respect, this kind of macroscopic measure can then serve as a guide for finding the correct theoretical model.

2
Models and methods

2.1 Statistical physics of complex systems

The beginning of statistical physics can be traced back to thermodynamics in the nineteenth century. The field is still very active today, with modern problems occurring in out-of-equilibrium systems. The first problems (up to c. 1850) were to describe the exchange of heat through work and to define concepts such as temperature and entropy. A little later many studies were devoted to understanding the link between a microscopic description of a system (in terms of atoms and molecules) and a macroscopic observation (e.g., the pressure or the volume of a system). The concepts of energy and entropy could then be made more precise, leading to an important formalization of the dynamics of systems and their equilibrium properties.

More recently, during the twentieth century, statistical physicists invested much time in understanding phase transitions. The typical example is a liquid that undergoes a liquid-to-solid transition when the temperature is lowered. This very common phenomenon turned out, however, to be quite complex to understand and to describe theoretically. Indeed, this type of "emergent behavior" is not easily predictable from the properties of the elementary constituents and as Anderson (1972) put it: "... the whole becomes not only more than but very different from the sum of its parts." In these studies, physicists understood that interactions play a critical role: without interactions there is usually no emergent behavior, since the new properties that appear at large scales result from the interactions between constituents. Even if the interaction is "simple," the emergent behavior might be hard to predict or describe. In addition, the emergent behavior depends, not on all the details describing the system, but rather on a small number of parameters that are actually relevant at large scales (see for example Goldenfeld 1992).

Statistical physics thus primarily deals with the link between microscopic rules and macroscopic emergent behavior and many techniques and concepts have been developed in order to understand this translation – among them the notion of relevant parameters, but also the idea that at each level of description of a system there is a specifically adapted set of tools and concepts. When describing a fluid, we do not need in general to describe all its molecules; instead, a coarse-grained description is enough to characterize many phenomena occurring in such systems. The importance of scales in the description of a system and the link between microscopic interactions and emergent collective behavior are two of the important reasons that pushed physicists to think that these tools and concepts could actually be "exported" to more complex socioeconomic systems, for example cities.

This line of thought brings us to the idea, discussed some time ago, that a city is an emergent phenomenon resulting from agents interacting with each other. This point of view is not in contradiction with the microeconomic view, which considers that macroscopic phenomena result from interactions and motivated actions of agents in an uncertain environment. As discussed above, we do know that in many systems only a very few ingredients are relevant at a large scale. The problem is then to correctly identify the minimal set of interactions and micromotives that is able to explain the observed emergent behavior.

This view of a city as a self-organized system was discussed by geographers (Pumain and Moriconi-Ebrard 1997; Pumain 2004; Batty 2007) and economists (Krugman 1996) and has direct implications on how we study this entity. Indeed, many urban studies were devoted to a specific area, and even though this is interesting and important from many points of view, it is however less relevant to the task of finding dominant mechanisms and parameters that govern city evolution. In this respect, we are more interested here in universal aspects of cities that can be found for a wide range of urban systems, irrespective of their different origins, locations, or histories. At the root of this is the belief that all cities are various expressions of the same "object." The same economic and social forces are in play, but lead to different cities depending on the history of the particular system. There is of course a strong path dependence for such a system; the simple expression of this is the fact that built areas and infrastructures are difficult (and costly) to destroy and to reconstruct differently. For these reasons, we will focus in this book on these generic aspects and we will not insist on specific cities, their history and their structure, on which better books have already been written.

2.2 The shape of a science of cities

In physics the description of a system usually depends on the scale at which the phenomenon is considered and also on the particular aspect we want to describe. The relationship between scale and the corresponding description of a system has been highlighted in statistical physics and could probably be applied to many complex systems. We can illustrate this with simple examples. If we are interested in writing the equation for a fluid traveling in a pipe, it is useless to use quantum mechanics in order to describe the fluid at a molecular level. A coarse-grained view is enough to describe the fluid and leads to the Navier–Stokes equation. If we are interested in the conductivity of a material, there is no immediate need to take into account other features such as its elasticity or bulk modulus (even if there is an obvious connection). In some sense the description and the corresponding mathematical framework are adapted to the observation scale and to the particular phenomenon that is under scrutiny. For a given scale there is however a set of fundamental principles. For instance, in the classical description of the fluid, we use Newton's equations, which are valid at a macroscopic scale and neglect all quantum (and relativistic) effects. Once we have the scale and the corresponding set of principles, we can go ahead and try to quantitatively characterize the process in which we are interested. This point of view should in principle be applicable to many complex systems and in particular to cities. The search for a "grand unified theory of cities" might be a little too ambitious, and also useless since each problem has its own scale and set of principles and each level might require a whole new conceptual structure.

The wording "Science of cities" is not very precise, but we can nevertheless try to propose what could constitute solid and useful foundations of such a science and get some inspiration from physics. The discussion of the interface between physics and cities is not unlike the current debate about a theoretical physics of biological systems and we may repeat here the arguments given by Bialek (2015). In order to construct a science of cities, we can indeed refer to physics that has a long history of describing phenomena in mathematical terms. Physics is not a catalogue of disparate models but rather a small set of mechanisms which, once applied to a particular system, have great predictive power. Very roughly, in physics, when we say that we understand a system it usually means that:

 (i) we know the dominant mechanisms;
 (ii) we are able to construct a minimal model, with predictions consistent with empirical observations and measures;
(iii) we can propose new experiments and applications, and how to modify the system in order to improve it.

In particular, from (i) we can predict the response of the system when we apply a perturbation to try to modify it. In the case of cities we could thus offer some general large-scale insights when facing different planning alternatives.

In physics, we thus understand a phenomenon when we have a set of tools such as mechanisms and equations that allow us to describe the system quantitatively. For the conductivity of materials, we know that the main mechanism is the movement of electrons subjected to an electrical field. From this we can then elaborate more complex theories, taking into account quantum effects, the presence of impurities, and so on. This knowledge then allows first-order prediction of the behavior of any material (usually things are not that simple, but nonetheless, they work like this in principle). This type of knowledge is probably what we need in urban science. Urban planning is usually about changing physical (infra)structures. Examples of important operations range from green belts, to new towns, to completely planned towns (such as Saudi Arabia's King Abdullah Economic City or Songdo in South Korea). Understanding the main forces and being able to predict the effect of various changes in an urban system would thus be an important goal for a science of cities.

2.3 How many parameters?
2.3.1 Statistical physics and relevant parameters

Statistical physics is primarily concerned with the passage from a microscale description of a system to its macroscopic behavior as described by a small set of variables. A particularly important case concerns phase transitions for which we observe drastic changes in the organization of a system depending on the value of some parameter (such as temperature). Many studies were devoted to this problem and it is impossible to describe here all the achievements and results that have been obtained regarding these phenomena. Techniques and concepts were developed in order to understand these systems, such as the idea of relevant parameters, or the renormalization group (see for example Stanley 1971; Goldenfeld 1992).

The idea that only a few parameters are actually relevant for describing the macroscopic behavior is probably the one that will prove most useful across other disciplines. Indeed, when we shift our focus on a physical system from a small to a large scale, the remarkable fact is that only a few parameters are relevant for the description of the system (this can be described quantitatively via the renormalization group; see for example Goldenfeld 1992). A gas, for example, is made of a very large number of atoms (on the order of 10^{23}), but in order to describe its macroscopic behavior only a few variables (temperature, volume, etc.) are needed. This dependence on a small number of relevant parameters is also at the heart of the "universality" observed in seemingly different systems which, however, in fact share the same small set of relevant variables. These relevant parameters include for example the dimension of the system, the dimension of local parameters, and usually the existence of correlations. Phenomena that have the same relevant parameters but are otherwise unrelated will then have the same critical behavior. Physicists indeed showed that the Ising model, for example, can describe the behavior at phase transitions of systems as diverse as magnetic materials, interacting neurons, and more recently social systems (Bouchaud 2013).

A very simple expression of this idea that many details actually do not matter too much can be found in the central limit theorem (see for example Feller's classical text of 1957 and Bouchaud and Georges' 1990 review). In its simplest form this famous theorem states that the sum S_N of N independent identically distributed random variables X_i,

$$S_N = \sum_{i=1}^{N} X_i \qquad (2.1)$$

has a distribution that can be approached (in a central region typically of order at least $N^{1/2}$; see for example the discussion on this point in Bouchaud and Potters (2003)) by a "universal" distribution that depends on the two first moments of X (when they exist; otherwise, there is a generalization of the central limit theorem, see Bouchaud and Georges 1990). All the details encoded in the higher moments of X become irrelevant as soon as N is large enough. We also see in this simple illustration the importance of correlations. In the simple example of the sum of random variables, the existence of correlations can indeed alter the validity of the central limit theorem, and more generally, correlations between the constituents of a system can lead to a nontrivial complex collective behavior that is difficult to predict.

2.3.2 Modeling cities

The state of the art for modeling urban systems can essentially be described by the graph sketched in Fig. 2.1. On one hand, models in urban economics have a small number of parameters, but on the other their link to empirical data is usually weak. The connection to real-world data is usually made through stylized fact, but "hard" tests such as comparing quantitative predictions with empirical data are still very sparse. A stylized fact – a term generally used in economics – refers to empirical findings that are consistent across a wide range of examples, time periods, etc. In economics, a theoretical model is usually validated when it can explain, qualitatively, these observed stylized facts. On the other hand, agent-based simulations such as land-use–transport integrated (LUTI) models, which describe the complex relationship between land use and transportation in detail, usually are very specific and integrate a large number of parameters. As such they can easily give the appearance of fitting the data and their validity is difficult to assess. In particular, it is unclear how much we can trust the predictions of these models when the system undergoes non-negligible, unexpected perturbations.

30   *Models and methods*

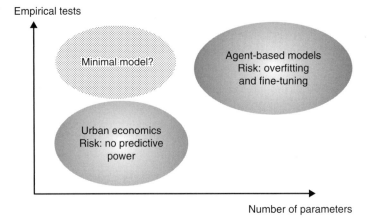

Figure 2.1. Urban economics proposes models with a small number of parameters; in contrast, agent-based models (such as LUTI models) rely on a very large number of parameters. Ideally, a minimal model has a small number of parameters and is able to predict many empirical findings.

The problem with parameters is actually not limited to the description of cities but occurs in the study of many complex systems (a prominent example being biological systems; see Bialek 2015). There are several problems with parameters and we discuss here two of them:

- The number of parameters is very large. This could mean that we are actually missing an additional level of unification that eliminates some of these (redundant) parameters.
- Reproducing a phenomenon requires fine tuning of parameters. This could mean that we are missing some dynamical mechanism that makes this apparent fine tuning occur naturally.

In physics, many successful theories are free from tunable parameters, and as noted in the case of biology by Bialek (2015), the "intrusion" of physicists into urban studies can be understood as a reaction against the proliferation of parameters. A model is satisfactory when the number of parameters is very small, and the number of correct predictions is very large.

The recent dramatic increase in availability of data on cities offers the possibility of convergence to such a minimal model. Once we understand and have exhibited the hierarchy of mechanisms governing the evolution of cities, we can propose a minimal model with a small number of parameters and with predictions that are testable against empirical data. Such a model, along the same lines as models developed in physics, could be expected to be sufficiently robust to be able to cope with and describe the behavior of the system in the event of large perturbations.

Physicists usually start with the simplest model with the smallest number of parameters and compare the outcome of such a simplistic model to empirical observations. This empirical feedback then pushes the theorist to correct the model by adding a parameter, or proposing an entirely new mechanism. This way of proceeding seems natural to many scientists but is at odds with urban modeling as done with LUTI models (see for example Acheampong and Silva's 2015 review) that incorporate a very large number of parameters from scratch, making it difficult to assess the relevance of each parameter – the model is a black box with a large number of parameters. It is then not really surprising when these models are sometimes able to reproduce empirical observations, as this is more a case of overfitting a phenomenon than explaining it.

In order to construct a model, we note that physicists never clung to their pre-existing tools but rather adapted their tools to the phenomenon studied. We observe that in urban economics, mathematical tools derived from the theory of optimization and more recently from optimal transport might actually hinder adaptation to the description of reality. An example discussed in this book (Section 5.2) is the classical model of job search (McCall 1970), which relies on optimal strategy. Despite the fact that the theory is very elegant, finding the optimal strategy is in general very difficult and it is hard to believe that the model describes the actual decision process experienced by individuals looking for a job.

2.4 Critiques of urban economics

The models developed in urban economics have produced a number of interesting predictions and allow the effect of various parameters to be discussed, but they suffer from a number of drawbacks. First, they almost inevitably assume that the spatial structure of the city is monocentric. Indeed in most models, the spatial variable is the distance to the center (usually the central business district). Although this seems to be a reasonable approximation for small cities where there is one center that concentrates almost all activities, it seems to become inaccurate as cities grow. In fact, we might wonder what is the dominant force that creates different secondary activity centers; we will discuss this in Chapter 3.

Second, all these models assume that the city is in equilibrium. In a system, when there are only two time scales, for fast (τ_{fast}) and slow (τ_{slow}) processes, we might consider that the system is indeed in equilibrium for times such that $\tau_{fast} \ll t \ll \tau_{slow}$. For real-world cities, however, there are many different time scales. Mobility has a time scale of hours, migration and demographic effects act on a basis of months or years, and infrastructure generally varies on much longer time scales. The relevance of equilibrium in this multiscale system is at least challenged and probably not realistic at all. In addition, as early as the 1970s some authors had discussed the possibility that the dynamics of cities is intermittent, with bursts of activity over relatively short periods of time.

Third, an important feature of cities is their path dependence or in other words the fact that they result from the superimposition of layers corresponding to different time scales. This is a particularly important issue as most models in urban economics do not consider this type of "irreversibility" in which individuals at time t have to cope with the current state of affairs and cannot start again from scratch.

Finally, as already mentioned, these models were thought of as theoretical guides but in general were not tested thoroughly against data, usually because of the lack of empirical results. The problem is indeed not the many simplifications (one-dimensional monocentric cities, identical agents, etc.) but how much we can trust these models and how they are related to reality. Model validation by data is usually questionable and the predictions of the model are generally not tested directly, but by statistical means only.

We next discuss in more detail some of the points addressed above, including the equilibrium assumption and also a more technical but nevertheless very important point: the problem of the choice of the utility function.

2.4.1 Interactions and equilibrium

Interactions

As discussed above, correlations are in general relevant effects. Their nature and presence usually alter emergent macroscopic behavior in a qualitative way. Correlations between the constituents of a system are thus particularly important but are very often neglected in classical models of urban economics, even though they appeared very early in other disciplines such as quantitative sociology (Schelling 1971). Indeed, most economic models consider a "typical" agent who maximizes his utility without considering what other agents are doing. Interactions can lead to a large variety of behaviors that can be extremely relevant for socioeconomic systems, such as path dependence, aging, or punctuated equilibria (Bouchaud 2013).

Equilibrium

A crucial assumption in urban economics is that of equilibrium, which is somehow surprising. Most of the systems encountered in nature are not in equilibrium. They might evolve over a time scale τ_{evol} that is much larger than the typical observation scale,

$$\tau_{evol} \gg \tau_{obs}, \tag{2.2}$$

and in this case, it is natural to consider the system in equilibrium at a time scale of order τ_{obs} (provided that τ_{obs} is much larger than the time scale of fast processes). For a city, the situation is obviously much more complex, because of the coexistence of a large number of processes and agents. In fact, we have a whole range of time scales, from smaller ones – related to daily mobility for example – to very large ones that describe the evolution of infrastructures. The most important time-varying quantity, the population, varies also significantly over different time scales. Birth and death rates produce a natural time scale, but another important source of population change in cities is migration, which can occur on smaller time scales. For example, we can estimate a few of these orders of magnitude for the United States:

- The average worker has approximately 10 jobs in a lifetime and moves about 12 times. This leads to approximately 6–7 years in a given location (data from the National Labour Survey).
- A typical growth rate of a percent per year for a city of about 1 million thus corresponds on average to 1,000 new inhabitants per month.
- Growth of area is obviously slower; to have an idea of the order of magnitude we can consider a typical case of a large city of 1,000 km^2 with a yearly urban land growth rate of order 1%. This corresponds to 10 km^2/yr; if we consider an isotropic growth, we then obtain an increase in city radius of order 10 m/yr. In reality this growth is certainly concentrated in areas with transport infrastructure and amenities, and is thus larger in these areas, but nonetheless, these orders of magnitude suggest that cities hardly behave like systems in equilibrium even in terms of built area.

In addition, from a theoretical point of view, the statistical physics of disordered systems (see for example Mézard et al. 1990) tells us that the relaxation time needed for a system composed of N interacting agents can in the worst case grow as

$$\tau_{\text{rel}} \sim e^N \tag{2.3}$$

because of the existence of significant barriers that are difficult to overcome. This result shows that under certain conditions (for a more complete discussion, see Bouchaud 2013), the socially optimal equilibrium will never be reached, because τ_{rel} grows exponentially with the number of individuals.

For all these reasons, it seems problematic to consider cities as actually systems in equilibrium states resulting from the optimization of some quantity. An important goal for a theory of cities would then be to describe the time evolution of space- and time-dependent population density $\rho(x,t)$. Such an equation would then inform us about the relevant time scales and the possibility of considering cities, in some regimes, as being in equilibrium.

2.4.2 Invariance with respect to utility choice

Another important problem is the choice of the utility function in models of urban economics. Indeed, if in some favorable cases the final outcome of the theory does not depend on the particular choice of the utility function, this is not always the case, leading to many issues. In order to illustrate this problem, we start with a "good" case where the general trend predicted by the theory is actually independent of the specific form of the utility function (provided that it satisfies some general reasonable constraints). We then continue this discussion with a specific case where the population density profile obtained from the model depends explicitly on the utility function.

A good case: the bid-rent gradient

We first illustrate this discussion through the classical Alonso–Muth–Mills (AMM) model (see Chapter 7 for more details). The problem consists of identical individuals in a monocentric city with a single transportation mode. These individuals are looking to rent accommodation of area $s(x)$ at location x and are subject to a budget constraint

$$Y = z + C_R(x) + T(x) \tag{2.4}$$

where Y is the individual's income, C_R and T are the rent and transportation costs, and z represents all other costs (also called composite commodities in economics). Here it is also assumed that there is a renting price profile per unit area $R(x)$ which depends on the location. The renting cost can then be written as $C_R(x) = R(x)s(x)$. The transportation cost is assumed to be a simple function of the distance to the CBD located at $x = 0$ and is written as $T(x) = V|x|/v$, where V is the value of time (see Section 7.3) and v the average velocity of the transportation mode. In urban economics, the standard assumption is that individuals optimize the utility which is a function of z and s only,

$$U = U(z, s) \tag{2.5}$$

subject to the budget constraint (see Chapter 7). This is a classical problem which can be solved with Lagrange multipliers:

$$\max_{z,s} U(z,s) \text{ subject to } Y = z + C_R + T$$
$$\Leftrightarrow \max_{z,s}[U(z,s) - \lambda(Y - z - C_R - T)] \tag{2.6}$$

where λ is a Lagrange multiplier. A faster way here is to introduce the constraint with $z = Y - C_R - T$ and to maximize $U(Y - R(x)s - T(x), s)$ with respect to s:

$$\frac{dU}{ds} = 0 = \partial_1 U(-R(x)) + \partial_2 U \tag{2.7}$$

where $\partial_i U$ denotes the derivative with respect to the ith variable. From this equation, we obtain the renting cost under the form

$$R(x) = \frac{\partial_2 U}{\partial_1 U}. \tag{2.8}$$

An additional requirement is that the maximum utility U^* should be independent of x. If it is not, then individuals could choose another, better location and we wouldn't be at equilibrium. We thus have to write

$$\frac{dU^*}{dx} = 0 = \partial_1 U \left(-s\frac{dR}{dx} - R\frac{ds}{dx} - \frac{V}{v}\right) + \partial_2 U \frac{\partial s}{\partial x} \tag{2.9}$$

where the functions $s(x)$ and $R(x)$ are computed at equilibrium. Combining Eqs. (2.8) and (2.9) which are valid for all x, we then obtain the central result for the Alonso–Muth–Mills (AMM) model (see for example Brueckner 1987):

$$\frac{dR}{dx} = -\frac{V}{vs(x)}. \tag{2.10}$$

We see from this last equation that the rent profile is a decreasing function with respect to the distance to the center, which signals the fact that in this model there is a trade-off between transportation and renting costs.

The remarkable point here is that this relation is valid for all utilities depending on z and s (subject to standard regularity and convexity conditions): we didn't have to specify the exact form of U. This result is rather strong: whatever the form of U (for reasonable choices) we obtain Eq. (2.10).

Problems arise: solving the model

The AMM model considered above gives a general form for the bid-rent gradient that is independent of the specific utility used. However, if we want to solve the model completely and compute the density and rent profiles, the specific form of the utility is crucial. We give the outline of the demonstration here; full details are given in Chapter 7.

In order to illustrate this, we first consider the one-dimensional model of a closed city with utility given by

$$U(z, s) = \alpha \log z + \beta \log s \tag{2.11}$$

where α and β are positive and can always be chosen such that $\alpha + \beta = 1$. This utility is an increasing function of both z and s, indicating the preferences for more space and more composite commodities. The prefactors α and β tune this preference for either more space or more commodities. The budget constraint is written as $Y = z + sR(x) + T(x)$. From this utility, we can express z as a function of s and u and obtain

$$Z(u, s) = e^{u/\alpha} s^{-\beta/\alpha}. \tag{2.12}$$

The bid-rent function Ψ is given by

$$\Psi(u, x) = \max_{s} \frac{Y - T(x) - Z(u, s)}{s}. \tag{2.13}$$

This maximum is obtained for $-s \, dZ/ds = Y - T - Z$, which after simple calculations leads to the following expression for the surface and the bid-rent functions:

$$s(u, x) = \alpha^{-\alpha/\beta} e^{u/\beta} [Y - T(x)]^{-\alpha/\beta} \tag{2.14}$$

$$\Psi(u, x) = \beta \alpha^{\alpha/\beta} e^{-u/\beta} [Y - T(x)]^{1/\beta} \tag{2.15}$$

At this point we have two unknowns: the maximum utility at equilibrium u^* and the size x_f of the city. In order to determine them we use the two constraints for a closed city with N individuals (and with absentee landlords)

$$\int_{-x_f}^{x_f} \frac{dx}{s(u^*, x)} = N \tag{2.16}$$

$$R(x_f) = \Psi(u^*, x_f) = R_A \tag{2.17}$$

where R_A is the rural land-rent value. Choosing for example $R_A = 0$ and $T(x) = a|x|$ (which are the standard choices for these models) we obtain for the land rent and the density ($\rho = 1/s$) profiles

$$R(x) = \frac{Na}{2} \left[1 - \frac{|x|}{x_f} \right]^{1/\beta} \tag{2.18}$$

$$\rho(x) = \frac{Na}{2Y\beta} \left[1 - \frac{|x|}{x_f} \right]^{\alpha/\beta} \tag{2.19}$$

where the size of the city is given by $x_f = Y/a$. This value corresponds to the maximum distance when the income is entirely converted into transport cost. The density (and the land rent) is thus decreasing according to some function that depends on β. People close to the center pay more for rent, less for transportation and have smaller apartments. Individuals living further away from the center trade commodity for space and have larger apartments. For example, for $\beta = 1/2$ we have a sort of parabolic shape and the density varies as $\rho(x) \sim (1 - |x|/x_f)^2$.

Another *a priori* reasonable choice for the utility is given by

$$U(z,s) = \alpha z + \beta \log s \tag{2.20}$$

where α and β are positive and can still be chosen such that $\alpha + \beta = 1$. As above this utility indicates a preference for larger apartments and more commodities. It can also be solved completely and we obtain

$$R(x) = \frac{Na}{2} e^{-a\frac{\alpha}{\beta}|x|} \tag{2.21}$$

$$\rho(x) = Na \frac{\alpha}{2\beta} e^{-a\frac{\alpha}{\beta}|x|} \tag{2.22}$$

(the boundary condition here is $R(\infty) = 0 = R_A$). We thus observe that the profiles are very different from Eq. (2.19), with an exponential decay in this case. Maybe more importantly, with the first utility choice, the result Eq. (2.19) indicates that there is one length scale given by the income and transportation cost $x_f = Y/a$, while with the second choice the city possesses a length scale of order $\beta/(\alpha a)$ which is independent from the income level Y. The difference between the two results is thus not only quantitative but also qualitative, with different variables and interpretations.

Even if this model is elementary and omits many relevant issues (such as, for example, heterogeneities of agents and their interactions), the results discussed here highlight a fundamental issue. Indeed, we can expect that a well-behaved theory predicts profiles that are independent of the utility choice. The important question is thus the following one. Does the utility have a specific form (and we cannot choose it as we want), or does the equilibrium formalization not hold? This is obviously a crucial question since the utility formalization is a pillar of urban economics, but unless we can bring empirical evidence to support it, its scientific validity can be challenged.

2.5 Data

The crucial game changer in research on cities is data. We now have a large amount of data related to almost any aspect of urban life. We have data at many different scales and about many different processes (see Fig. 2.2). At short time scales, we have for example mobility data coming from cell phones that inform us about the location of individuals at every moment. Until recently this type of data was usually obtained by surveys that were very limited in time and space, while in contrast cell-phone or GPS data give a precise, real-time picture of mobility in urban areas. In addition, the adoption of RFIDs in subway, bus, and streetcar

38 — Models and methods

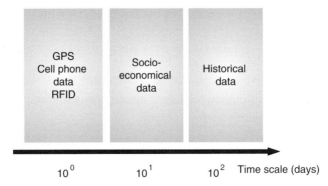

Figure 2.2. Schematic representation of the availability of different types of urban data and their corresponding time scales.

transport systems enrich this type of dataset and the increasing number of sensors and capture devices in cities that measure variables such as air pollution will allow us to make progress in our undersanding and modeling of mobility.

At the larger scale of months up to a year, we have socioeconomic data on various features such as the relation between income and space, the space-time evolution of real-estate prices, and so on. Also at the aggregate level of the city we can measure a number of variables that characterize the efficiency as well as the problems in a city and how they vary with population (see Section 7.4 on scaling in urban systems).

Finally, at a very long time scale, the digitization of historical documents allows monitoring of the long-term evolution of infrastructure. This type of information is crucial in the modeling of cities and the identification of the dominant forces that drive their evolution (see Chapter 4).

2.5.1 Sources

The availability of data about cities is ever increasing. Various agencies, state-owned and private companies release vast amounts of data at various spatial and temporal scales. The sources of these different data are usually new technologies such as smartphones and other personal devices.

A very important factor for the progress of science (and not only in the study of cities) is the availability of data. Provided that there are no privacy problems, all scientists should be able to download these datasets. This ensures the reproducibility of the results but also speeds the research process. Fortunately, this point of view is now shared with a growing proportion of scientific journals and will very soon be the standard. Another very important issue concerns the accuracy of these new datasets. It needs to be tested and also to be compared

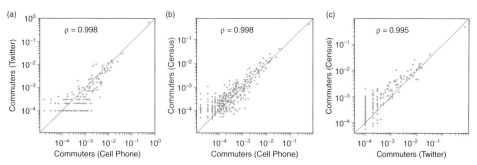

Figure 2.3. Correlations between different data sources: Comparison between the non-zero commuting flows obtained with three different datasets for the Barcelona's case study (the values have been normalized by the total number of commuters for both OD tables). Points represent each pair of municipalities and the straight line represents the $x = y$ line. (a) Twitter and cell phone. (b) Census and cell phone. (c) Census and Twitter. Figure from Lenormand et al. (2014).

with more traditional ways of obtaining socioeconomic data, such as surveys. Lenormand et al. (2014) investigated this issue for the origin-destination (OD) matrix (see Chapter 5) in the case of Spain for data obtained with Twitter, cell phones, and standard census. They obtained the results shown in Fig. 2.3 and the consistency between different datasets demonstrates their validity. This is obviously an encouraging result and further studies are probably needed in order to assess the quality of these new data sources. It also demonstrates the importance of having a variety of different data sources, permitting cross-checking of results.

For cities, there is the additional problem of achieving harmonized data and consistent definitions of cities. As discussed in Chapter 1, there is no consensus so far on the definition of urban areas (besides the obsolete administrative definitions), and this obviously renders comparisons among cities and the search for "universal" properties more difficult.

2.5.2 Different types of data

Network data

Infrastructures and the OD matrix can be characterized as networks. Concerning transportation networks, a large number are available on the web. In particular, street and road networks are available thanks to the Open Street Map initiative (www.openstreetmap.org). The cleaned and topologically corrected networks are freely available under a large number of formats.

More recently, with the development of studies on coupled networks (or multilayer networks) the need for such datasets increased a lot. Ideally, network data with the corresponding traffic volume would be the best for most studies. Unfortunately, traffic data are either unavailable or very difficult to find (when they exist). In some cases, however, we can bypass this problem and take advantage of other freely available datasets and use a reasonable proxy for traffic (such as the betweenness centrality for example; see Section 4.1) in order to enable the assessment of the structural efficiency of the system. This is the case for example for the set of timetables for all transportation modes in the UK (see Chapter 6 for a detailed discussion on multimodality). This allows to identify key quantities to characterize this coupled system, and to provide some directions in order to improve urban public transport.

For this multilayer system, in each layer the nodes represent stop areas. These nodes are connecting nodes between layers if different transportation modes stop in the same area (see details in Chapter 6 and in Gallotti and Barthelemy 2015). For example, the multilayer system has been constructed in detail for three megacities (Paris, Tokyo, and NYC; see Gallotti et al. 2016b). The data used to describe the lines in the metropolitan networks as of 2009 were mainly extracted from Wikipedia (www.wikipedia.com; see Roth et al. 2012) and yields a weighted spatial network: each station has geographical coordinates, and the edges connect consecutive stations on a metro line and have a weight corresponding to the trip duration along this edge. The connection between nodes in different layers can then be made according to a simple criterion. Indeed, by studying for example for Paris the cumulative distribution for the walking distances of these connections data, it was found that the 99th percentile corresponds roughly to a walking distance of order $d_w = 250$ meters. Motivated by this, we can then allow a maximum walking distance of $d_w = 250$ meters for the coarse-graining procedure that is needed to construct the multilayer networks in New York City (NYC) and Tokyo.

Geohistorical data; Temporal network data

The digitization of old documents such as maps opens the possibility to investigate the long time evolution of urban systems. This digitization of old maps, together with GIS (Geographical Information Systems) tools such as georeferencing, map matching, and so on, allows construction of "pictures" of urban systems, enabling their evolution to be reconstructed. These datasets are generally difficult to obtain and are usually produced thanks to collaborative work with tools such as the "Building Inspector," developed by the New York Public Library (http://buildinginspector.nypl.org), and others developed in France in the case of the Cassini map (see Perret et al. 2015 and www.geohistoricaldata.com).

In Strano et al. (2012), the authors constructed the evolution of a region located north of Milan (Italy) from various maps described in Table 2.1. These different historical maps describe the system at different times (namely at $t = 1833, 1914, 1933, 1955, 1980, 1994, 2007$) and give a single snapshot of the street network for

Table 2.1. *List of geographical information sources used to construct the dataset for the evolution of the Groane region studied in Strano et al. (2012). IMGI denotes here the Italian Military Geographic Institute and LR the Lombardy Region.*

Date	Source	Owner	Format
1833	Topographical Map of Kingdom of Lombardy–Venetia	IMGI	Raster
1914	Map of Italy	IMGI	Raster
1933	Map of Italy	IMGI	Raster
1955	Aerial Photography Survey	IMGI	Raster
1980	Lombardy Regional Map	LR	Raster
1994	Lombardy Regional Map	LR	Raster
2007	Mosaic of Urban Municipalities Plans	LR	Vectorial

each of these years. For each time frame, the primal graph was constructed (for the primal graph the junctions represent the nodes and the roads are the links of the networks). For maps, we have also the position associated with each element (for example the nodes) and an important problem in GIS is then to match different maps. In particular, if we have two graphs representing the same system at different times t and t', the challenge is to compare them and to decide whether two nodes from different time frames actually describe the same object (Costes et al.). There are also problems such as "reincarnations," where an object is represented at times t and $t'' > t$ but not at an intermediate time such that $t < t' < t''$. From the point of view of maps, we then see an object that disappears at time t' but reappears at time t''. More generally, the difficulty in these problems is to find and propose a map that is as close a representation as possible to the reality.

Other datasets of the same type were developed and used in other studies. The example shown in Fig. 2.4 displays the central part of the city of Paris (France) evolving from the Revolution until the present day (Barthelemy et al. 2013).

Mobility data

An important phenomenon that is discussed in Chapters 5 and 6 is individual mobility in urban areas. The structure of cities results from, and to a large extent conditions, mobility, which is a fundamental process for their evolution (Makse et al. 1995; Bettencourt 2013; Louf and Barthelemy 2014a). In addition, understanding mobility is also fundamental for a large variety of complex processes such as traffic forecasting (Axhausen and Gärling 1992), or epidemic spreading (Colizza et al. 2007; Balcan et al. 2009).

Until recently, most mobility data have been extracted from surveys with their inherent limitations: the number of individuals is usually not very large,

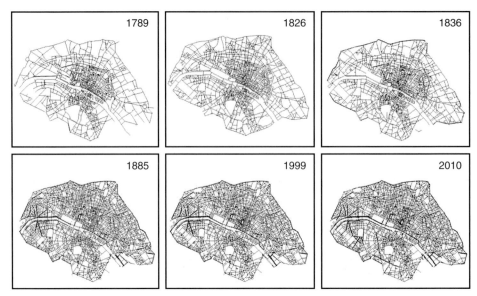

Figure 2.4. Evolution of central Paris from 1789 to today. Figure from Barthelemy et al. (2013).

and more importantly the accuracy of information such as commuting travel time and distance is usually low. With recent developments in information and communication technologies, the investigative focus shifted from these traditional surveys to several new data sources. In particular, it has become possible to follow individual trajectories from cell-phone calls (Gonzalez et al. 2008; Song et al. 2010a; Kang et al. 2012), location-sharing services (Cheng et al. 2011; Noulas et al. 2012; Liu et al. 2014), and microblogging (Hawelka et al. 2014); or directly extracted from public transport ticketing systems (Roth et al. 2011), GPS tracks of taxis (Liang et al. 2012; Liu et al. 2012; Wang et al. 2015; Liu et al. 2015; Tang et al. 2015), private cars (Bazzani et al. 2010; Gallotti et al. 2012), or single individuals (Rhee et al. 2011; Zhao et al. 2015). For most data sources, the spatial positions are the most reliable values and this information can be used for studying how far and where humans are moving.

We note that cell-phone data can actually be used for more than studying mobility and we refer the interested reader to the reviews by Blondel et al. (2015) and by Calabrese et al. (2015).

Socioeconomic data

The convergence of different datasets, such as administrative surveys or tax data, is also an important source. The gross domestic product (the "gross metropolitan product" for cities), the number of patents, and so on for different cities provide important indications of their economic activity and productivity. These data usually come from administrative compilations and tax documents that can nowadays be cross-referenced. However, the difficulty with this type of data is restricted access. These datasets contain personal information and access is usually limited to statistical institutes. Partial aggregations of some of these data are however available from national agencies such as the Census Bureau in the United States and other similar agencies in other countries.

2.5.3 Data are not enough: models

Models are essential guides for understanding the reality of systems and the processes that govern them. Data and empirical observations put constraints on the model and help in identifying the most important parameters and processes. Both these ingredients, data and models, are necessary in order to make progress and the success of physics lies precisely in the possibility of performing this feedback loop. If one of these ingredients is missing, the resulting explanation is questionable. This issue is particularly relevant in studies of cities in urban economics that consist often in untested laws and theories (Bouchaud 2008). Without empirical validation, it is not clear what urban economic models teach us about cities.

On the other hand, empirical measures can be misleading without a theoretical framework for understanding and interpreting them. An interesting and important example concerns the scaling (see Pumain 2004 and Section 7.4) of the quantity Q_{CO_2} of CO_2 emitted by cars in a city of population P, which follows the law

$$Q_{CO_2} \sim P^\beta \tag{2.23}$$

where the exponent β is positive. The fact that β is not what is naïvely expected on purely dimensional grounds, for example, signals important interactions and processes in this phenomenon.

This CO_2 problem is particularly timely: pollution peaks occur in large cities worldwide with seemingly increasing frequency, and are suspected to be the source of serious health problems (Bernstein et al. 2004). Many studies (Glaeser and Kahn 2010; Rybski et al. 2013; Fragkias et al. 2013; Oliveira et al. 2014) are interested in how CO_2 emissions scale with the population size of cities. The question they ask is simple: Are larger cities greener – in the sense that there are fewer emissions per capita for larger cities – or smoggier? Surprisingly, these different studies reach contradictory conclusions and we can identify two main sources of errors, which can be related to a lack of theoretical guidance (Louf and Barthelemy 2014b).

The first error concerns the estimate of the quantity of CO_2 emissions due to transportation. In the absence of a direct measure, Glaeser and Kahn (2010) based their estimate on the total commuting distance, but this is incorrect as in heavily congested areas time and distance behave very differently. Even if it is not the perfect indicator, the time spent to go from one point to another seems to be a better measure of the CO_2 emitted compared to the distance, which is independent of traffic and leads to a serious underestimation of CO_2 emissions.

The second important issue lies in the definition of the city itself (see Chapter 1 for a discussion of this issue), and thus over which geographical area the quantities Q_{CO_2} and P should be aggregated. There is currently much confusion in the literature about how cities should be defined, and there is no clear consensus yet. This is a crucial issue, as scaling exponents in general seem to be very sensitive to the particular definition of a city (Arcaute et al. 2013). In fact, aggregating over urban areas (UAs) or metropolitan statistical areas (MSAs) entails radically different behaviors (see Fig. 2.5). For the United States, using the definition of UAs provided by the US Census Bureau (2014), CO_2 emissions per capita sharply

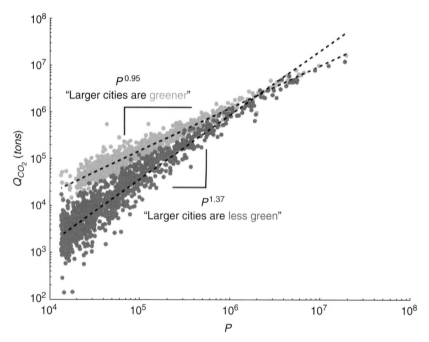

Figure 2.5. Are larger cities greener? Scaling of transport-related CO_2 emissions with the population size for US cities from the same dataset but at different aggregation levels. In dark gray, the aggregation is done at the level of Urban Areas and in light gray, for Combined Statistical Areas. Depending on how the city is defined, the scaling exponents are qualitatively different, leading to two opposite conclusions. Data on CO_2 emissions were obtained from the Vulcan Project (The Vulcan Project 2014; as in Fragkias, 2013, Oliveira, 2014). Data on Urban Areas and MSA populations were obtained from the Census Bureau (US Census Bureau, 2014). Figure taken from Louf and Barthelemy (2014b).

increase with population size, implying that "larger cities are less green." Using the definition of MSAs, also provided by the Census Bureau, one finds that CO_2 emissions per capita slightly decrease with population size, implying that "larger cities are greener." In the face of these two opposite results, it is difficult to draw any conclusions, but several questions arise. We can indeed question the empirical result: there is some noise in the data and the nonlinearity observed here could be an artefact due to it (see Leitao et al, 2016). This is related to the more important problem: the absence of any mechanistic insights about this scaling. This opens the door to misguided interpretations, and more importantly formulating policies based on such questionable results might have disastrous consequences. This example clearly illustrates the dangers of interpreting data without a theoretical guide and in the absence of the feedback loop between theory and data.

2.6 The barriers to interdisciplinarity

Cities are of interest to many people, not only in academia but also stakeholders, planners, and obviously, citizens. The number of questions and problems related to urban systems is thus vast and ranges from psychology, sociology, and health, to more mathematical fields such as geography, urban economics, transport, applied mathematics, and statistical physics. It seems surprising, however, to view these different elements in isolation: transport has to do with location theory, geography and urban economics discuss spatial distributions, and so on. This specialization into different fields is unhelpful for understanding cities and can sometimes lead to misunderstanding and to difficult communication.

For example, a quantity relevant to one field is completely inadequate for another one. This happens in particular between social scientists interested in qualitative approaches to cities and quantitative researchers. Average density is certainly not enough to describe the full complexity of a city, but it allows to distinguish between different types of organizational patterns. A typical misunderstanding among social scientists is the idea that quantitative researchers want to explain everything about cities. This is usually not the case: most models try to explain a particular aspect of cities. Social science is able to grasp the broader dimensions of cities, while quantitative scientists will insist on empirical observations and theoretical predictions. In this respect, the current tension between urbanism and quantitative approaches (Dupuy and Benguigui 2015) cries out for a dialog.

At the other end of the spectrum, urban economics had to develop with scarce datasets and did not insist on the quantitative validation of theoretical approaches. Also, concepts such as equilibrium, which are strongly connected to a specific framework and tools, might actually not be well adapted to these systems. Plasticity is required in the study of such a complex object and we have to be ready to abandon our favorite tool if it is simply not adapted to the object of study. Also, misunderstanding can arise due to a lack of dialog. Physicists certainly do not think that economics is easy (see for example Buchanan's answer, 2013, to economist Chris House) but their tools and their mathematical descriptions of reality could in some cases be helpful and should not be rejected on the basis that physicists are outsiders.

Finally, when a problem treated a long time ago in a specific discipline is tackled by scientists from another one, a number of problems can result. This has been recently discussed by O'Sullivan and Manson (2015) in the case of the "intrusion" of physicists into geography. When scientists from a discipline branch out to other ones, a number of traps exist. Besides the possibility of counterproductive arrogance, a lack of knowledge of the existing literature and the danger of reinventing the wheel are key factors. In particular, commonalities should be avoided and discussion should target important problems and seek concrete answers. However, the back-to-basics approach can sometimes lead to new insights and should not always be rejected. This is another problem for interdisciplinary approaches: the use of authority arguments for rejecting a result obtained by an "outsider." As O'Sullivan and Manson (2015) conclude, it is incumbent on the field with the prior claim to avoid narrow-minded defensiveness. An important step may then be accomplished, allowing scientists from different fields to work collaboratively and publish together.

There are therefore several barriers to interdisciplinary research, from both sides. The outsider should definitely know the subject and the literature of the domain, but experts in that domain should not reject the outsider too hastily, and should consider the results on a purely scientific basis. Only dialog can be productive here, and it is probably a necessary condition for a successful construction of a science of cities.

3
The spatial organization of cities

The locations of homes, activities, and businesses shape a city, and identifying the mechanisms that govern these spatial distributions is crucial for our understanding of these systems. We present here some recently discussed aspects, which may provide a basis for further insights. We will begin with a discussion on the location of stores and facilities, which are very likely governed by optimal considerations.

We will then discuss the polycentric aspects of cities by starting with their identification and measures. We will describe how to characterize and measure an activity center – a "hotspot" – defined as a local maximum of the activity density. The important empirical result is that the number of these hotspots scales sublinearly with the population size.

We continue by describing two classical theoretical models for polycentricity: the Fujita–Ogawa model, proposed in the 1980s, which relies on the idea that agglomeration effects are responsible for polycentricity, and the edge-city model proposed by Krugman. As we shall see, these models cannot, however, explain the scaling of the number of hotspots with population and this leads us to reconsider the classical Fujita–Ogawa model in order to derive a result in agreement with empirical observations.

3.1 Optimal locations

3.1.1 Distribution of public facilities

Public facilities such as airports, post offices, and hospitals have to be distributed according to the local population density in order to optimize their efficiency. These facilities constitute an important part of the urban structure and help to shape the spatial distribution of population. It is therefore important to understand the organization of these particular places.

We can measure these spatial distributions, and the natural null model to compare against these empirical observations is the optimal case where the average distance from an individual to the nearest facility is minimized (Gastner and Newman 2006), and we follow here the derivation given by Gusein-Zade (1982). We assume that the population is distributed in a domain D according to a space-dependent local density $\rho(x)$ and we examine the distribution of N facilities such that their locations $x_1, x_2, \ldots, x_N$ minimize the total distance F to reach them,

$$F(x_1, x_2, \ldots, x_N) = \int \rho(x) \min_{i=1,\ldots,N} |x - x_i| \mathrm{d}x. \tag{3.1}$$

These N facilities partition the area under consideration into Voronoi cells. The Voronoi cell for a facility i is the set of points that are closer to this facility than others. We denote by $a(x)$ the area of the Voronoi cell to which the point x belongs. The distance to the facility for these individuals will then be of order $g\sqrt{a(x)}$, where g is a geometrical factor of order 1 that depends on the shape of the Voronoi cell. The function to be minimized can then be rewritten as

$$F[a(x)] \sim \int \rho(x)\sqrt{a(x)}\mathrm{d}^2 x \tag{3.2}$$

$$\text{such that} \int_D \frac{1}{a(x)} \mathrm{d}^2 x = N \tag{3.3}$$

where the constraint expresses that we have N facilities. The functional derivative then reads

$$\frac{\delta}{\delta a(x)} \left[\int \rho(x)\sqrt{a(x)}\mathrm{d}^2 x + \lambda \left(\int \frac{1}{a(x)} \mathrm{d}^2 x - N \right) \right] = 0 \tag{3.4}$$

where λ is the Lagrange multiplier. This equation leads to

$$\frac{\rho(x)}{\sqrt{a(x)}} = \frac{2\lambda}{a(x)^2} \tag{3.5}$$

which gives $a(x) \sim (\lambda/\rho(x))^{2/3}$. Inserting this result in the constraint allows us to extract λ and we get

$$\rho_f(x) - \frac{1}{a(x)} = C\rho(x)^{2/3} \tag{3.6}$$

where ρ_f is the density of facilities and where the prefactor is $C = N / \int \rho(x)^{2/3} \mathrm{d}^2 x$.

As noted by Gusein-Zade (1982), this simple argument has a number of consequences on the statistics of the Voronoi tesselation. In particular, we have the following properties:

3.1 Optimal locations

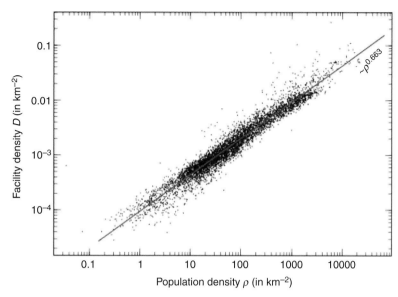

Figure 3.1. Density of facilities versus population density ρ shown here for the US (in log-log). The power-law fit gives a slope of 0.663 ± 0.002 in agreement with the theoretical analysis. Figure taken from Gastner and Newman (2006a).

- The area of the Voronoi cells scales as

$$a(x) \sim \rho(x)^{-2/3} \qquad (3.7)$$

- The average radius of Voronoi cells scales as

$$r(x) \sim \rho(x)^{-1/3} \qquad (3.8)$$

- Finally, the number of individuals in each Voronoi cell is given by

$$n(x) \sim \rho(x)a(x) \sim \rho(x)^{1/3}. \qquad (3.9)$$

The result Eq. (3.6) was tested by Gastner and Newman (2006), who plotted the facility density ρ_f versus the population density ρ for 5,000 facilities (hospitals, airports, or malls) for the 48 lower states of the United States; their results are shown in Fig. 3.1. The density of facilities is thus scaling with the density of individuals raised to a power consistent with $2/3$. This result suggests that (at least for the US) the spatial distribution of facilities indeed satisfies an optimality argument.

3.1.2 Distribution of retail stores

Facilities are obviously important, but the spatial organization of retail commercial activities is also a crucial component in the spatial organization of a city. Different retail stores can attract or repel each other depending on their activities. This interaction dictates the spatial arrangement of these retail stores, and Jensen (2006) proposed an empirical study of this problem for the city of Lyon, France.

In order to characterize the interaction between two types of activities, Jensen (2006) used the "M" index defined by Marcon and Puech (2009) (see also Section 7.2). The interaction between two categories of stores, A and B, is estimated in the following way. We assume that in the city, we have a total number N of stores and $N_{A(B)}$ stores of category $A(B)$. For each store of category A (located at position x), we draw a circle of radius r around it and in this circle we count the total number n_{tot} of stores and the number of stores of category B. We then compute the following ratio for this store at location x:

$$M_{AB}(x,r) = \frac{\frac{n_B(x,r)}{n_{\text{tot}}(x,r)}}{\frac{N_B}{N_{\text{tot}}}}. \tag{3.10}$$

This ratio thus compares the local density of stores of category B to the uniform case. If this ratio is significantly larger than one, there is an attraction between A and B and a repulsion in the opposite case. Note that in some cases, the denominator can be chosen to be $n_B/(n_{\text{tot}} - n_A)$, which takes into account the fact that there could be a strong attraction between A stores, leading to an apparent repulsion between A and B.

We then average over all A stores and obtain the M-index between A and B:

$$M_{AB}(r) = \frac{1}{N_A} \sum_{x \in A} \frac{n_B(x,r)}{n_{\text{tot}}(x,r)} \Big/ \frac{N_B}{N_{\text{tot}}}. \tag{3.11}$$

Jensen (2006) took a constant value of $r = 100$ m for the radius, which corresponds to a typical distance that a potential customer accepts to visit different stores. However the variation of this M-index with r is also interesting and contains useful information (see Marcon and Puech 2009 for a detailed discussion of this point).

From this quantity, we can define $a_{AB} = \log M_{AB}$, which is positive if there is attraction and negative otherwise. This quantity echoes the coupling between magnetic spins, a very important model in the study of phase transitions (see, for example, Goldenfeld, 1992). Different directions are then possible, and the first interesting idea is to maximize the total satisfaction defined by

$$K = \sum_{i,j} a_{\sigma_i \sigma_j} \pi_{\sigma_i \sigma_j} \tag{3.12}$$

where $\pi_{\sigma_i \sigma_j} = 1$ if $\sigma_i = \sigma_j$ and $\pi_{\sigma_i \sigma_j} = -1$ if $\sigma_i \neq \sigma_j$. The quantity i represents possible locations for stores and σ_i is the category of store at location i. Jensen

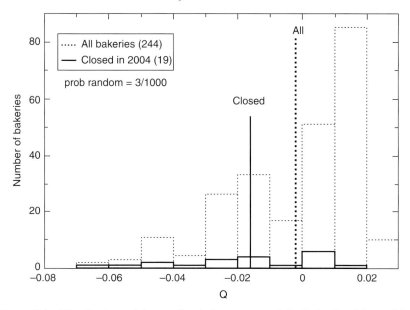

Figure 3.2. Distribution of the quality index computed for bakeries. Most of the bakeries that closed between 2003 and 2005 had on average a lower quality index compared to the whole set of bakeries. Figure taken from Jensen (2006).

(2006) obtains five homogeneous groups of retail stores, which achieves a good satisfaction for most elements of the groups. Remarkably, this classification, which emerges from an optimization calculation, matches the one given by the US Deparment of Labor Standard Industrial Classification System.

Another possibility consists in using these interaction coefficients in order to characterize quantitatively good locations for a store of a given category. The idea proposed by Jensen (2006) is to compute (in a radius r) the number $n_{ij}(x)$ of neighbors of type j around a site of category i located at x. The "quality" of the environment centered at x is then given by

$$Q_i(x) = \sum_j a_{ij} \left[n_{ij}(x) - \overline{n_{ij}} \right] \qquad (3.13)$$

where $\overline{n_{ij}}$ is the average over all locations x. A large index thus corresponds to an environment in which the "favorite" neighbors of i are over-represented. In Fig. 3.2 we show the distribution of this quality index computed for bakeries. Nineteen bakeries closed in the period 2003–2005 and their average quality index was indeed much lower than the average quality index computed for all bakeries ($2.2 \times 10^{-2} \ll 4.6 \times 10^{-3}$).

This type of approach is remarkable not only in the sense that it opens interesting research directions, but also because it connects standard concepts and tools

3.2 Measuring a polycentric structure

3.2.1 Definition

The density of activity is not in general uniform in space and a way to study this distribution is to identify local maxima. In many cases, space is discretized, and the data gives access to the spatial density $\rho(i)$ of employees (or any other quantity describing activity), where i denotes a cell in the discretization grid (in some cases, we have $\rho(i, t)$ at different times of day). This density distribution is a complex entity and we have to extract relevant and useful information. The locations that display a density much larger than others – the hotspots – give a good picture of the city by showing the concentration points and how they are organized in space.

We have in general at least two spatial scales in this problem. The smallest scale is given by the size of the unit (denoted by a), which is given by the data. It could be for example the typical size given by census tracts, or in the cell-phone data case, the typical size given by the antenna interspacing. In the search for local maxima we also need a "resolution parameter" $r \geq a$ that is used for smoothing the surface obtained by the raw data defined on geographical units. Another interpretation for this scale is related to what we actually call a hotspot. In Fig. 3.3 we represent schematically what we observe depending on the resolution parameter. At the scale of buildings for example, each store or office is a local maximum and we observe

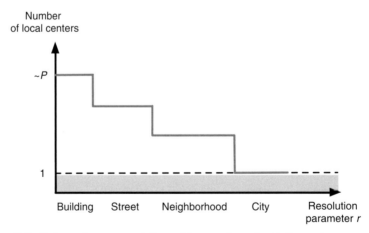

Figure 3.3. Schematic representation of the number of activity centers versus the resolution parameter. At a very small scale, local maxima correspond to stores, at a slightly larger scale, we observe commercial streets, at an intermediate scale neighborhoods, and at a very large scale, we see the city as a single activity center.

Figure 3.4. Employment density in two US cities (darker areas for greater employment density). Left: San Antonio, TX, a large city with a complex structure. Right: A smaller city (Winter Haven, FL), with a clear monocentric organization. Figure courtesy of R. Louf.

a number of hotspots roughly proportional to the population (or at least a very large number). If we smooth out over a scale larger than buildings we observe hotspots that correspond to commercial streets, for example. At an even larger scale, local maxima correspond to neighborhoods, and eventually we observe a single activity center that corresponds to the city as a whole. The identification process of hotspots consists then in two steps. The first one is the determination of local maxima (at the scale a given by elementary units). Once we have these "elementary" local maxima, we aggregate them over a scale given by the resolution parameter r (which corresponds to the smoothing of the density surface). The notion of a hotspot will then depend on the scale, and in what follows we will consider a resolution of order a kilometer or less, which means that a hotspot is a neighborhood or an area where all points are within walking distance.

The determination of centers and subcenters is a problem which has been widely tackled in urban economics (Giuliano and Small 1991; McMillen 2001; McMillen and Smith 2003). Starting from a spatial distribution of densities, we thus have to identify the local maxima. For example in Fig. 3.4, we see that there are indeed local maxima and that their number and organization contain important information about the structure of these cities. This is in principle a simple problem solved with the choice of a threshold δ for the density ρ: a cell i is a hotspot if the density satisfies

$$\rho(i) > \delta. \tag{3.14}$$

This is for example what was done by Giuliano and Small (1991) to determine employment centers in Los Angeles. It is however clear that this method introduces some arbitrariness due to the choice of δ, and also requires prior knowledge of the city to which it is applied to choose a relevant value of δ. Nonparametric methods have also been applied to determine the number of centers, some based on the regression of the logarithm of employment density versus the distance from the center (McMillen 2001), others on the exponent of the negative exponential fit of the density distribution (Griffith 1981). These methods are however limited by the fact that the actual density distribution cannot be properly fitted by an exponential law. In the next section, we discuss an alternative method that circumvents this problem and allows us to control the impact of the choice of a threshold value.

3.2.2 Identifying and counting hotspots

We present here a nonparametric method proposed by Louail et al. (2014) to determine hotspots. For the sake of simplicity we assume that the density distribution $\rho(i)$ is known for each unit i of the city (if the data for this city are decomposed in spatial units based on either a grid or another spatial division). The method is however very general and can be used for any quantity depending on space.

A first simple criterion is to choose for the threshold the average $m = \overline{\rho(i)}$ of the distribution: all the cells whose density is larger than m are hotspots. This is indeed a "weak" definition of what can be considered as a hotspot, but it can be used as a "lower" bound $\delta_{\min} = m$. In order to understand how the various properties of hotspots will depend on the choice of a threshold we have to define an upper bound for the threshold. In order to characterize the disparity of the activity in the city and to isolate the dominant places, we first plot the Lorenz curve of the density distribution. The Lorenz curve, a standard object in economics, is a graphical representation of an empirical probability distribution. In order to construct this plot, we sort the densities $\rho(i)$ by increasing rank, and denote them by $\rho(1) < \rho(2) < ... < \rho(n)$, where n is the number of cells. The Lorenz curve is constructed by plotting on the x-axis the proportion of cells $F = i/n$ and on the y-axis the corresponding proportion L of the density given by

$$L(i) = \frac{\sum_{j=1}^{i} \rho(j)}{\sum_{j=1}^{n} \rho(j)}. \tag{3.15}$$

The Lorenz curve is then defined by $L = L(F)$. If all densities are of the same order the Lorenz curve is close to the diagonal from $(0, 0)$ to $(1, 1)$. In general we observe a convex curve (see Fig. 3.5) with a more or less strong curvature, and the area between the diagonal and the actual curve is related to the Gini coefficient, an important indicator of inequality used in economics.

The stronger the curvature of the Lorenz curve, the stronger the inequality in the data, and intuitively the smaller the number of hotspots. We can then propose a

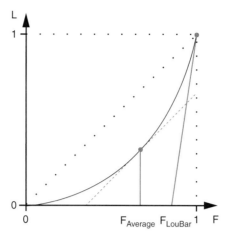

Figure 3.5. Illustration of the criteria selection on the Lorenz curve. The intersection of the slope at $F = 1$ with the x-axis gives the threshold F_{Loubar}, while the point with slope equal to one (dashed line) corresponds to the average density (and value F_{Average}). Figure from Louail et al. (2014).

criterion by relating the number of hotspots (those cells that have a very high value compared to the others) to the slope of the Lorenz curve at the point $F = 1$: the larger the slope, the smaller the number of dominant individuals in the statistical distribution. A natural way to identify the typical scale of the number of hotspots is then to take the intersection point F_{Loubar} between the tangent of $L(F)$ at point $F = 1$ and the horizontal axis $L = 0$ (see Fig. 3.5). The stronger the curvature, the larger the slope at $F = 1$ and consequently the larger F_{Loubar}. A larger F_{Loubar} indicates a stronger constraint to be a hotspot, in agreement with the previous discussion. We note here that the average criterion F_{Average} corresponds to the point of the Lorenz curve with slope equal to 1. Indeed, the general expression of the Lorenz curve for the set of densities $\rho(i)$ whose cumulative function is $F(\rho)$ is

$$L(F) = \frac{1}{m} \int_0^F \rho(F) \mathrm{d}F \qquad (3.16)$$

where $\rho(F)$ is the inverse function of the cumulative (we can check on this expression that $d^2L/dF^2 > 0$). This point F_{Average} satisfies, by definition,

$$\frac{\mathrm{d}L}{\mathrm{d}F} = 1 \qquad (3.17)$$

and gives $m = \rho(F_{\text{Average}})$; in other words, the hotspots will be those with densities larger than the average. In contrast, the more restrictive criterion based on the slope at $F=1$ gives

$$F_{\text{Loubar}} = 1 - \frac{m}{\rho_M} \quad (3.18)$$

where ρ_M is the maximum value of $\rho(i)$. We thus see that in general $F_{\text{Average}} < F_{\text{Loubar}}$ and that this new criterion, more restrictive, does not only depend on the average value of the density but also on the dispersion: as ρ_M increases, the value of F_{Loubar} increases and therefore the number of detected hotspots decreases.

All other possible and reasonable methods will then give a value contained in the interval $[F_{\text{Average}}, F_{\text{Loubar}}]$ between the average criterion and the new criterion. Instead of choosing a particular point, we can then study most of the properties for hotspots computed with a threshold in the range $[F_{\text{Average}}, F_{\text{Loubar}}]$. In particular, this will enable us to test the robustness of results against the arbitrariness of the hotspot identification.

Once we have determined the hotspots we can study various properties such as their spatial arrangements, their evolution in time (see for example Louf and Barthelemy 2013; Louail et al. 2014). The simplest property is the number of hotspots and from employment data in the US (at the census tract scale), we observe that the number h of employment hotspots scales with the city population P as

$$h \sim P^{\beta} \quad (3.19)$$

where $\beta \approx 0.64$ for the US case. Using cell-phone data for estimating the activity in largest urban areas in Spain (Louail et al. 2014), we observe a similar behavior with the number of centers scaling with $\beta \approx 0.46$ for working hotspots and $\beta \approx 0.60$ for residential hotspots (see Fig. 3.6) computed from the activity and residential

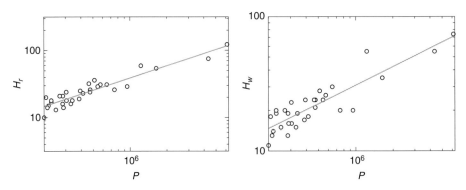

Figure 3.6. Number of residential (left) and work (right) hotspots for 31 urban areas in Spain. The power-law fit gives fo residences $\beta \approx 0.60$ ($r^2 = 0.93$), and for workplaces $\beta \approx 0.46$ ($r^2 = 0.89$). For more details, see Louail et al. (2014).

densities, respectively. We note here that the results are robust for various levels of aggregation from 500 m to 2 km.

Remarkably enough, the values of the exponent β are of the same order for the US employment data and the Spanish cell-phone data. These results are important as they confirm the polycentric structure of cities, but also represent a valuable guide for testing theoretical approaches. The number of hotspots scales sublinearly with the population and this fact should be explained by any model that aims to explain polycentricity.

3.3 Polycentricity: Classical approaches

We review here two important theoretical approaches that provide explanations for the polycentric organization of cities: the Fujita–Ogawa model (Fujita and Ogawa 1982) and the edge-city model (Krugman 1996).

3.3.1 The Fujita–Ogawa model

The Fujita–Ogawa model (Fujita and Ogawa 1982) is an interesting model that discusses the location choice for both individuals and firms. In the next section, we will simplify this model and extract testable predictions; here we present its original version.

The purpose of this model is to provide an explanation for the fact that – at equilibrium – the monocentric organization is not always stable and that a polycentric structure can be observed. The main ingredient here is the agglomeration effect of companies, signaled by an attraction of large-density regions for location choice.

More precisely, the model describes the equilibrium of both individuals who are looking for a job and a residence and firms that have to optimize their profit. Individuals (or households) optimize their utility $U(S, Z)$ which depends on both the land consumption S and the composite commodity Z. This quantity Z represents the amount of money left after having paid the rent and transportation costs:

$$Z = W(y) - R(x) - T(x, y) \qquad (3.20)$$

where x denotes the residence of the individual and y the location of her job. The quantity $W(y)$ is the wage profile, $R(x)$ the renting cost profile, and $T(x, y)$ the transport cost from x to y. In their approach, Fujita and Ogawa neglected congestion and wrote for the transport cost

$$T(x, y) = td(x, y) \qquad (3.21)$$

where $d(x, y)$ is the (Euclidean) distance from x to y. If we assume that the rented area is fixed (equal to S_h) for households, the utility optimization is equivalent to maximizing Z given by Eq. (3.20) (in the following we will also assume the same for firms with value S_b). If we also assume that all households are identical, they should achieve the same utility level U^* at equilibrium and hence the same value for Z^*.

The firms on the other hand are optimizing their profit,

$$\Pi(y) = F(y) - R(y) - L(y)W(y) \qquad (3.22)$$

where Π is the profit for a company located at y, $K(y)$ the benefit for coming to location y, and $L(y)$ the number of individuals working at location y. The term $L(y)W(y)$ is thus the payroll at location y. The classical assumption is that $L(y)$ is constant and given by L_0 so that there is a simple relation between the number of workers N and the number of firms M

$$N = ML_0 \qquad (3.23)$$

(assuming that all individuals are working). The important term in the firm profit is the benefit or "locational potential" F and the central point in the Fujita–Ogawa paper is the discussion of the agglomeration effect described by

$$F(y) = \int K(y - y')b(y')dy' \qquad (3.24)$$

where $b(x)$ is the density of firms and where $K(y)$ is a kernel with typical spatial extension $1/\alpha$ (which means that it falls to zero over distances larger than $1/\alpha$). The standard choice in many studies is to take

$$K(y) = ke^{-\alpha|y|}. \qquad (3.25)$$

Similarly to households, we assume that all business firms are identical, and the profit level Π^* must then be the same at equilibrium, regardless of location.

In order to discuss the equilibrium properties of this model, it is convenient (Ogawa and Fujita 1980) to introduce the fraction of households (described by their density $h(x)$) that are commuting to location y,

$$P(x, y) = \frac{\text{\# commuters } x \to y}{\text{\# residents at } x}. \qquad (3.26)$$

Another (equivalent) quantity is the "commuting function" $J(x)$, which gives the commuting location for an individual living at location x. In the original paper and in the following we will consider the one-dimensional case (the 2-D case was adressed by Ogawa and Fujita 1989, leading to similar results). At this point, we thus have the following set of unknowns:

3.3 Polycentricity: Classical approaches

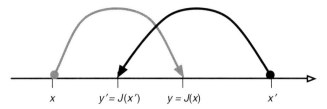

Figure 3.7. Example of cross-commuting that is not allowed at equilibrium in the Fujita–Ogawa model.

- the household and business firm density functions $h(x)$, and $b(x)$,
- the land rent profile $R(x)$,
- the wage profile $W(x)$,
- the commuting pattern $P(x, y)$ (or equivalently the commuting function $J(x)$), and
- the utility and profit level U^* and Π^* at equilibrium.

The first important result is about cross-commuting (see Fig. 3.7). If we consider an individual at a fixed location x, this worker at equilibrium will commute to $y = J(x)$ such that the disposable income for land and composite commodities is a maximum. In other words, at equilibrium, the commuting function $J(x)$ is such that

$$W(J(x)) - t|x - J(x)| = \max_{y} \left[W(y) - t|x - y| \right]. \tag{3.27}$$

This condition leads to the impossibility of cross-commuting such as represented by Fig. 3.7. Indeed, let's assume, without loss of generality, that $W(J(x)) > W(J(x'))$. It is then clear that we have $W(y') - t|x - y'| > W(J(x)) - t|x - J(x)|$ so that $J(x)$ has to be equal to y'. Depending on the distances, we can then have $J(x') = y$ or $J(x) = y'$ but in both cases, no cross-commuting is possible.

If we now assume two (ordered) locations $0 \leq x \leq x'$, non-cross-commuting then implies $J(x) \leq J(x')$ (or the opposite if these quantities are negative). We denote $y = J(x)$ and $y' = J(x')$, and we assume that $x \leq x' \leq y \leq y'$ – all the other cases can easily be worked out. The individual located at x is commuting to y and not to y', and we therefore necessarily have

$$W(y) - t(y - x) \geq W(y') - t(y' - x). \tag{3.28}$$

Similarly the individual located at x' commutes to y' and not to y, and therefore

$$W(y') - t(y' - x') \geq W(y) - t(y - x'). \tag{3.29}$$

Eq. (3.28) leads to $W(y) - W(y') \geq t(y - y')$ and Eq. (3.29) leads to $W(y') - W(y) \geq t(y' - y)$. Both these conditions together give $W(y') - W(y) = t(y' - y)$

and by taking the limit $x \to x'$ we obtain

$$\frac{dW(y)}{dy} = t. \quad (3.30)$$

Similarly, in the case where $y < y' < x < x'$, it is easy to show that we have

$$\frac{dW(y)}{dy} = -t. \quad (3.31)$$

In order to state the equilibrium conditions for this model, we can define – as is usual in these problems (see Chapter 7) – the bid-rent function $\Psi(x)$ for households and the maximum land rent $\Phi(x)$ a firm could pay at location x and for a profit level π:

$$\begin{cases} \Psi(x) = \max_y \{W(y) - Z(U) - td(x, y) | U(Z) = U\}, \\ \Phi(x) = kF(x) - \pi - W(x)L_0. \end{cases} \quad (3.32)$$

Following Fujita and Ogawa, we can define different area types such as the exclusive residential area (RA), the exclusive business area (BD) and the integrated district (ID):

$$RA = \{x \mid h(x) > 0,\ b(x) = 0\} \quad (3.33)$$

$$BD = \{x \mid h(x) = 0,\ b(x) > 0\} \quad (3.34)$$

$$ID = \{x \mid h(x) > 0,\ b(x) > 0\}. \quad (3.35)$$

We also assume that the city is surrounded by rural land with rent R_A. The necessary and sufficient conditions for the system defined by the set of variables $\{h(x), b(x), R(x), W(x), P(x, y), U^*\}$ are then (see Chapter 7):

i. $R(x) = \max\{\Psi^*(x), \Phi^*(x), R_A\}$. In particular, in residential areas,

$$R(x) = \Psi^*(x) \quad \text{(RA)} \quad (3.36)$$

$$R(x) = \Phi^*(x) \quad \text{(BD)}; \quad (3.37)$$

ii. $R(x) = R_A$ on the urban fringe.

In addition to these equilibrium conditions, we also have a certain number of constraints (in addition to trivial non-negativity constraints):

- normalization:

$$\int h(x)dx = N \quad (3.38)$$

$$\int b(x)dx = M = N/L_0; \quad (3.39)$$

3.3 Polycentricity: Classical approaches

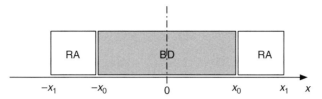

Figure 3.8. Monocentric geometry considered by Fujita and Ogawa (1982). The center is the business district (BD) and is surrounded by residential areas (RA).

- density constraint:

$$S_h h(x) + S_b b(x) \leq 1; \quad (3.40)$$

- labor market condition:

$$L_0 b(y) = \int P(x,y) h(x) \mathrm{d}x. \quad (3.41)$$

If we accept the idea of a city as a system in equilibrium, this model is of course very attractive and seems to capture the most important ingredients for understanding the spatial organization of firms and residences. As often happens with very general descriptions, it is unfortunately impossible to solve this model and Fujita and Ogawa (1982) essentially considered the stability of simple cases. They started with the stability of the monocentric situation depicted in Fig. 3.8, where a symmetric, one-dimensional city is organized around a central business district in area $[-x_0, x_0]$ and residential areas in $[x_0, x_1]$ and $[-x_1, -x_0]$. The densities are thus given by

$$h(x) = 1/S_h, \quad b(x) = 0 \quad \text{for } x \in \text{RA} \quad (3.42)$$
$$h(x) = 0, \quad b(x) = 1/S_b \quad \text{for } x \in \text{BD} \quad (3.43)$$

and the normalizations then impose

$$x_0 = \frac{S_b N}{2L_0}, \quad x_1 = \frac{S_b + S_h L_0}{2L_0} N. \quad (3.44)$$

As discussed above, the non-cross-commuting condition (Eq. (3.31)) leads to $dW/dx = -t$, which implies

$$W(x) = W(0) - tx. \quad (3.45)$$

The locational potential is

$$F(x) = \int_{-x_0}^{x_0} b(y) e^{-\alpha|x-y|} dy$$

$$= \frac{1}{\alpha S_b} \left(2 - e^{-\alpha(x_0+x)} - e^{-\alpha(x_0-x)}\right) \text{ for } x \in [-x_0, x_0] \quad (3.46)$$

$$= \frac{1}{\alpha S_b} \left(e^{-\alpha(x-x_0)} - e^{-\alpha(x+x_0)}\right) \text{ for } x \in [x_0, \infty]. \quad (3.47)$$

We can now solve the problem completely by using $\Psi^*(x) = (W(x) - Z^*)/S_h$, $\Phi^*(x) = (kF(x) - W(x)L_0)/S_b$ together with boundary conditions, but in order to discuss the stability of this configuration we note that the equilibrium conditions read

$$R(x) = \Phi^*(x) \geq \Psi^*(x) \text{ for } x \in [0, x_0]$$

$$R(x) = \Phi^*(x) = \Psi^*(x) \text{ at } x = x_0$$

$$R(x) = \Psi^*(x) \geq \Phi^*(x) \text{ for } x \in [x_0, x_1]$$

$$R(x) = \Phi^*(x) = R_A \text{ for } x = x_1$$

and since $\Psi(x)$ is linear and $\Phi(x)$ concave, a necessary condition for these relations to hold is that $\Phi^*(0) - \Phi^*(x_0) \geq \Psi^*(0) - \Psi^*(x_0)$, which after some simple manipulations leads to

$$\frac{t}{k} \leq \min\left\{\frac{F(0) - F(x_0)}{x_0}, C\frac{F(x_0) - F(x_1)}{x_1 - x_0}\right\}, \quad (3.48)$$

where $C = S_h/(S_b + S_h L_0)$. This is the important result here and we now discuss it. The order of magnitude of the right-hand side is given by α and the monocentric situation thus becomes unstable if t/k becomes larger than α or, in other words, if transport cost becomes too large compared to the typical inverse interaction distance between firms. Also, we can anticipate that if there is congestion, it amounts to a larger effective transport cost $t = t(N)$, reinforcing the instability of the monocentric organization for large cities, as discussed in Section 3.4.

In their paper Fujita and Ogawa (1982) test the stability of other situations, and obtain stability conditions for each case. This however shows the limitations of this approach: it is possible to test the stability of a given spatial organization, but we don't know to what it will converge and, for a general value of the population, what will be the typical organization. For example, it seems extremely difficult to evaluate here the number of activity centers (as defined by clusters of firms for

3.3.2 The edge-city model

This model was discussed by Krugman (1996) and is a good example of a minimal model that could probably be built upon and would be amenable to predictions that can be tested. The most important aspect here, as in the Fujita–Ogawa model, is the presence of interactions between firms. These interactions can lead to a polycentric organization of the city in which businesses are concentrated in spatially separated clusters. We follow the discussion proposed by Krugman and consider that the city is one-dimensional and the density of businesses is described by the function $\rho(x,t)$, which can vary in time and whose spatial integral is assumed to be constant. We assume that all locations are initially equivalent (which means in particular that there is a relatively uniform transport infrastructure) and that the attractiveness of a location will depend on the spatial distribution of businesses. In order to describe this mathematically, Krugman introduces a quantity $\Pi(x,t)$, a "market potential" which describes the level of attractivity of location x and is similar to the locational potential introduced in the Fujita–Ogawa model. Very generally, one can assume that there are both positive and negative effects due to business concentration and that these effects decay with distance. More precisely, we write

$$\Pi(x,t) = \int K(x-z)\rho(z,t)dz \qquad (3.49)$$

where the kernel is chosen as $K(x) = A(x) - B(x)$ with functions A and B that are both decreasing with distance. These functions represent positive and negative spillovers and how they vary with distance. Businesses will have an incentive to come to certain locations depending on the level of attractiveness Π, and the simplest assumption is to compare $\Pi(x,t)$ with the average spatial level $\overline{\Pi}(t) = \int \Pi(x,t)\rho(x,t)dx$:

$$\frac{d\rho(x)}{dt} = \gamma \left[\Pi(x,t) - \overline{\Pi}\right]. \qquad (3.50)$$

The density will then increase at locations where $\Pi(x,t) > \overline{\Pi}$ and decreases otherwise. This equation describes in a very simple way the evolution of the business density with time and provides some explanation for the self-organized nature of cities. Numerical simulations indicate that this nonlinear system indeed leads to various solutions with multiple centers at different locations, depending on the initial conditions. Essentially, if positive spillovers are larger we will observe bigger clusters of businesses. We also see that this concentration has a reinforcement effect: regions with a large market potential will be more attractive and will therefore grow.

In order to continue the discussion at a more quantitative level, we denote by λ the size of spatial fluctuations and by r_1, r_2 the ranges of positive and negative spillovers, respectively. For large frequencies, $\lambda \ll r_1, r_2$ there is basically a compensation effect of positive and negative spillover and we do not expect the growth rate to be large. Also, for very low-frequency fluctuations such as a maximum of $\rho(x)$ which decays slowly around x, and for strong enough negative spillover, the growth rate at x will be negative. Both low and high frequencies have negligible growth rate and it is thus natural to expect a frequency with a maximal growth rate, well-tuned to the spatial decay of positive and negative spillovers, leading to a specific spatial pattern in this city. In order to get a quick analytical insight, we can linearize the equation for $\rho(x)$ around the flat city $\rho(x) = \rho_0 + \delta\rho(x)$ and find

$$\frac{d\delta\rho}{dt} \simeq \gamma \int K(x-z)\delta\rho(z)dz \qquad (3.51)$$

where we choose the normalization $\int \rho(x) = 1$. Using the Fourier transform

$$\delta\rho(k) = \int e^{ikx}\delta\rho(x)dx \qquad (3.52)$$

we obtain, by integrating the linear differential equation (3.51),

$$\delta\rho(k) \sim e^{\gamma \hat{K}(k)t} \qquad (3.53)$$

where $\hat{K}(k)$ is the Fourier transform of the kernel. This expression thus shows that the Fourier mode k^* for which $\hat{K}(k)$ is maximum will develop faster and will lead to the appearance of a spatial pattern characterized by k^*. We can obtain explicit expressions with the choice $K(x) = Ae^{-|x|/r_1} - Be^{-|x|/r_2}$, which leads to

$$\frac{d\delta\rho(k,t)}{dt} = 2\gamma \left[\frac{Ar_1}{1+(r_1k)^2} - \frac{Br_2}{1+(r_2k)^2} \right] \delta\rho(k,t). \qquad (3.54)$$

We thus have for each mode a growth rate proportional to

$$\Lambda(k) = \frac{Ar_1}{1+(r_1k)^2} - \frac{Br_2}{1+(r_2k)^2} \qquad (3.55)$$

which satistifes $\Lambda(0) = Ar_1 - Br_2$ (which is negative for large enough B) and $\Lambda(k) \to 0$ for k large. We thus expect in general a maximum Λ^* for a value k^* that can be easily computed. This value k^* depends here on r_1, r_2, A, and B and is thus finite and independent from the city size. For a one-dimensional city of size L, the number of business activities is then given by

$$h \sim Lk^* \qquad (3.56)$$

which predicts a linear increase of the number of hotspots with city size, but doesn't explain how this quantity scales with population (in the two-dimensional case we would obtain $h \sim Ak^{*2}$). Although it is interesting to see how simple nonlinear effects give rise to a nontrivial spatial pattern, this model doesn't produce at this stage predictions that are directly testable against empirical data.

3.4 Revisiting the Fujita–Ogawa model

As described above, we observe empirically for various data that the number of activity centers scale with the population as

$$h \sim P^\beta \qquad (3.57)$$

where β is less than one and in fact lies in the range $[0.5, 0.7]$. This is a relatively simple fact that calls for a theoretical explanation. The Fujita–Ogawa and edge-city models provide some possible qualitative explanations but do not predict such a nonlinear scaling. There are not many simple empirical results about polycentricity and it thus seems reasonable to require from a theory that it explains, and in the best case reproduces, the scaling Eq. (3.57) with an exponent $\beta < 1$.

In order to illustrate this point, we discuss here a simplified version of the Fujita–Ogawa model (Louf and Barthelemy 2013, 2014a). The main point is to show that by omitting certain details and focusing instead on basic processes we can reach some sensible conclusions. We therefore aim at building a minimal model which captures the complexity of the system and reproduces the observed scaling. The model that we discuss here is in essence dynamical and describes the evolution of cities' organization as their population increases. We focus on car congestion – mainly due to journey-to-work commutes – and its effect on choice of job location for individuals.

3.4.1 A complex quantity described as random

We begin with the assumption that mobility patterns are mostly driven by the daily commuting and we would like to understand how an individual, given her household location, will choose her job location. As we will see in Chapter 5, on mobility, this process is at the heart of our understanding of the structure of cities. We assume that this choice will be determined by two dominant factors: the expected wage at a given job, and the commuting time to this job's location. Indeed, places with high average salaries are attractive, but having to spend a significant amount of time commuting every day is less desirable. We assume that there are potential activity centers $j = 1, \ldots, N_c$ in the city, each characterized by an average wage $w(j)$ at location j. This wage is *a priori* endogenously determined and depends *a priori* on many factors such as agglomeration effects, the type of industry, and so on. However, as we saw above, writing the full Fujita–Ogawa model leads to equations that are not tractable and cannot bring any clear testable predictions. A similar situation arises in physics when one studies the behavior of atoms comprising a large number of electrons. Physicists discovered (Dyson 1962) that instead of writing the exact equations, a statistical description of these systems relying on random matrices could lead to predictions in agreement with experimental results. The idea, then, is to replace a complex quantity such as wages – which depends on so many factors and interactions – by a random one. In this way we treat the wage as if it was exogenous and random, and write

$$w(j) = s\,\eta_j \tag{3.58}$$

where s represents the typical income in this city and η is a random number chosen uniformly in $[0, 1]$ (other choices for this distribution have no impact on the final results). Furthermore, we assume that the commuting time between i and j does not depend on the distance only, but also on the traffic T_{ij} between these two locations. We thus write a generalized cost for transport of the simplified form

$$T(i, j) = V \frac{d_{ij}}{v} \left[1 + \left(\frac{T(j)}{c} \right)^\mu \right] \tag{3.59}$$

where V is the value of time and v the average velocity of cars in this city. The quantity d_{ij} is the Euclidean distance between i and j (both supposed to be scattered randomly across the city), $T(j)$ the total incoming traffic at j, c the capacity of the underlying transportation network, and μ is an exponent describing the sensitivity of the network to congestion. The exponent μ is usually larger than one (Branston 1976). An individual living at i will thus commute to the center j which corresponds to the best trade-off between income and commuting time, thus to the center j such that the quantity

$$Z_{ij} = \eta_j - \frac{d_{ij}}{\ell} \left[1 + \left(\frac{T(j)}{c} \right)^\mu \right] \tag{3.60}$$

3.4 Revisiting the Fujita–Ogawa model

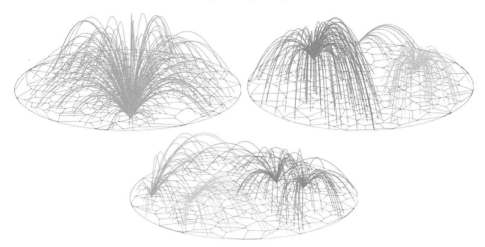

Figure 3.9. The monocentric (top left), distance-driven polycentric (top right), and attractivity-driven polycentric (top right) regimes as produced by our model. Each link represents a commuting journey to an activity center. Figure from Louf and Barthelemy, 2014a.

is maximum. The quantity ℓ is the maximum distance that people can financially travel daily, defined as the ratio of typical individual income and transportation costs per unit of distance.

3.4.2 Monocentric-polycentric transition

Depending on the relative importance of wages, distance, and congestion, the model defined by Eq. (3.60) predicts the existence of three different regimes (see Fig. 3.9): the monocentric regime the distance-driven polycentric regime, and the attractivity-driven polycentric regime.

Starting from a small city with a monocentric organization, traffic is negligible and $Z_{ij} \approx \eta_j$, which implies that all individuals are going to choose the most attractive center (with the largest value of η_j, say η_1). As the number P of households increases, however, traffic will also increase and some initially less attractive centers (with smaller values of η) might become more attractive, leading to the appearance of new activity centers, or "subcenters," characterized by non-zero numbers of commuters. More precisely, a new subcenter j will appear when for an individual i, we have $Z_{ij} > Z_{i1}$. Traffic so far is described by $T(1) = P$ and $T(j) = 0$, which leads to the equation for the appearance of a new center:

$$\eta_j - \frac{d_{ij}}{\ell} > \eta_1 - \frac{d_{i1}}{\ell}\left[1 + \left(\frac{P}{c}\right)^{\mu}\right]. \tag{3.61}$$

We assume that there are no spatial correlations in the subcenter distribution, so that we can make the approximation $d_{ij} \sim d_{i1} \sim L$ where L is the typical size of the city. The new subcenter will thus be such that $\eta_1 - \eta_j$ is minimum, implying that it will have the second-largest value denoted by $\eta_j = \eta_2$. For a uniform distribution, on average $\overline{\eta_1 - \eta_2} \simeq 1/N_c$ and we obtain from Eq. (3.61) a critical value for the population:

$$P^* = c \left(\frac{\ell}{L N_c} \right)^{1/\mu}. \tag{3.62}$$

Whatever the system considered, there is therefore always a critical value for the population above which the city becomes polycentric. We note that the smaller the value of μ (or the larger the value of the capacity c), the larger the critical population value P^*, meaning that cities with a good road system capable of absorbing large traffic display a monocentric structure for a longer period of time.

The monocentric regime is therefore fundamentally unstable with regards to population increase, which is in agreement with the fact that no major city in the world exhibits a monocentric structure. This is also in agreement with the general result of Fujita and Ogawa, but the transition described here is of a different nature, triggered by congestion.

3.4.3 Number of centers

Having established that cities will eventually adopt a polycentric structure, we might wonder how the number of subcenters varies with the population. We compute the value of the population at which the hth center appears. We still assume that we are in the attractivity-driven regime and that, so far, $h - 1$ centers have emerged, characterized by $\eta_1 \geq \eta_2 \geq ... \geq \eta_{h-1}$, with numbers of commuters $T(1), T(2), ..., T(h-1)$, respectively. The next worker i will choose a new center h if

$$Z_{ih} > \max_{j \in [1, h-1]} Z_{ij} \tag{3.63}$$

which reads

$$\eta_h - \frac{d_{ih}}{\ell} > \max_{j \in [1, h-1]} \left\{ \eta_j - \frac{d_{ij}}{\ell} \left[1 + \left(\frac{T(j)}{c} \right)^{\mu} \right] \right\}. \tag{3.64}$$

As can be checked on numerical simulations, the distribution of traffic $T(j)$ is narrow, which means that all the centers have roughly the same number of commuters $T(j) \sim P/(h-1)$. As above we also assume that the distance between the workers' households and the activity centers is typically $d_{ij} \sim d_{ih} \sim L$. The

3.4 Revisiting the Fujita–Ogawa model

previous expression (Eq. 3.64) now reads

$$\frac{L}{\ell}\left(\frac{P}{(h-1)c}\right)^{\mu} > \max_{j\in[1,h-1]}(\eta_j) - \eta_h. \qquad (3.65)$$

Following our definitions, $\max_{j\in[1,h-1]}(\eta_j) = \eta_1$, and according to order statistics, if the η_j are uniformly distributed, we have on average $\overline{\eta_1 - \eta_h} = (h-1)/(N_c+1)$. It follows from these assumptions that (i) the hth center to appear is the hth most attractive one, and (ii) the average value of the population $\overline{P}_h$ at which the hth center appears is given by

$$\overline{P}_h = P^*(h-1)^{\frac{\mu+1}{\mu}}. \qquad (3.66)$$

Conversely, the number h of subcenters scales sublinearly with population as

$$h \sim \left(\frac{P}{P^*}\right)^{\frac{\mu}{\mu+1}}. \qquad (3.67)$$

It is interesting to note that this result is robust: the dependence is sublinear, *whatever the distribution* of the random variable η. We can therefore conclude that, probably very generally and under mild assumptions, the number of activity subcenters in urban areas scales sublinearly with population, with the prefactor and the exponent depending on the properties of the transportation network.

While agglomeration economies seem to be the basic process explaining the existence of cities, this model brings evidence that congestion is a driving force that tears them apart. The non-trivial spatial patterns observed in large cities can thus be understood as a result of the interplay between these competing processes. This model could represent an important step towards a quantitative, predictive science of cities. More generally, this microscale approach is an interesting example of an out-of-equilibrium model: it is governed by local optimization with saturation effects, leads to different regimes, and is characterized by non-trivial dynamical exponents. It integrates economical ingredients in a single dynamical framework, unifies mobility patterns, the spatial structure of cities, and scaling. It could serve as a basis on which we could elaborate further by integrating information about firm locations, the influence of public transport, the integration of cities into larger systems (Rozenblat and Pumain 2007), and so on.

In the following we derive some of the consequences of this model on mobility in cities and associated quantities such as CO_2 emissions due to transport.

3.4.4 Consequences for mobility

Area

At this stage, the number of centers is a function of population and the area,

$$h = F(A, P) \qquad (3.68)$$

and we need an additional equation in order to get a closed system. We investigate here two different approaches. It is worth noting that both approaches give results in qualitative agreement, showing that some stylized facts – such as super- or sublinearity – are very robust.

Fitting procedure. In the absence of knowledge of the processes responsible for urban sprawl, we assume that the area behaves as

$$A \sim P^b \qquad (3.69)$$

where b is the exponent to be determined, through fits on data. The empirical value for the exponent for the US data is $b \simeq 0.85$. Once this exponent is given we then compute the various exponent for many quantities of interest. We get, for the number of centers h (using Eq. 3.69 and Eqs. 3.62 to 3.67),

$$h \sim P^{\frac{\mu+b/2}{\mu+1}} \qquad (3.70)$$

which is sublinear as long as $b < 2$, in agreement with the empirical results for US cities. As we will see, this approach yields the same qualitative behavior as that predicted with the method of the next section.

Coherent growth. We assume here that the scaling of the area A with population is determined by the number of activity centers and a constant commuting length of individuals. This means that the growth of the city's area is controlled by the number of new activity centers. If we assume that a city is organized around h activity centers and that the attraction basins of each of these centers are spatially separated, we then have $A \sim h A_1$, where A_1 is the area of each subcenter's attraction basin. This area A_1 is related to the average individual commuting distance L_{tot} by $\sqrt{A_1} \sim L_{\text{tot}}/P$, and we obtain

$$A \sim h \left(\frac{L_{\text{tot}}}{P}\right)^2 = h \ell_c^2. \qquad (3.71)$$

This leads to an expression for the number of centers:

$$h \sim P^{\frac{2\mu}{2\mu+1}} \qquad (3.72)$$

which is always smaller than 1, also in agreement with the empirical results for US cities. We can now also compute the area scaling:

$$\frac{A}{\ell_c^2} \sim \left(\frac{P}{c}\right)^{\frac{2\mu}{2\mu+1}}. \tag{3.73}$$

We further assume that L_{tot}/P is a fraction of the longest possible journey ℓ individuals can afford, that is to say

$$\ell_c \sim \ell. \tag{3.74}$$

The final expression for the area is then given by

$$\frac{A}{\ell^2} \sim \left(\frac{P}{c}\right)^{2\delta} \tag{3.75}$$

where $\delta = \frac{\mu}{2\mu+1}$. The exponent δ is smaller than $1/2$ for any $\mu \geq 0$, which implies that the density of cities increases *sublinearly* with population. In other words, the density of cities *increases* with population. We verify this with data on land area of urbanized areas in the US, where we find $2\delta_{\text{emp}} = 0.85 \pm 0.01$ (95% CI) which is not too far from the theoretical value $2\delta_{\text{th}} = 0.64 \pm 0.12$ (95% CI).

Because the area of an urban system results from centuries of evolution, we do not *a priori* expect this simple model – where individual vehicles are assumed to be the only vector of mobility – to give a prediction valid for all countries and all times. Nevertheless, these results give reason to believe that the spatial structure of the journey-to-work commuting should still be the dominant factor in the dependence of land area on population.

Total commuting distance

Using $L_{\text{tot}} \sim P$ and Eq. (3.75) we are now able to compute $L_{\text{tot}}/\sqrt{A}$, obtaining

$$\frac{L_{tot}}{\sqrt{A}} = P\left(\frac{P}{c}\right)^{-\delta}. \tag{3.76}$$

We can now estimate the exponent of the scaling of L_{tot} with the population. We find $L_{\text{tot}} \sim P^{0.68}$, which agrees very well with the value 0.66 measured by Samaniego and Moses (2008) directly from the data on total miles driven daily in more than 400 cities in the US.

Total road length

If we use the previously derived expression for the area A, we find for $L_N \sim \sqrt{AP}$

$$L_N \sim \ell \sqrt{P} \left(\frac{P}{c}\right)^{\delta}. \tag{3.77}$$

The quantity δ is less than $1/2$, which implies that L_N scales sublinearly with the city's population size. In other words, larger cities need less road per capita than smaller ones: we recover the fact that agglomeration of people in urban centers involves economies of scale for infrastructures. Within the fitting assumption (Eq. (3.69)), we would obtain $(1+b)/2$, leading to the same qualitative conclusions.

Total delay due to congestion

Unfortunately, agglomeration in cities does not only generate economies. Congestion, for instance, is a major diseconomy associated with the concentration of people in a given area. A simple way to quantify the impairment caused by traffic congestion is through the total delay it generates. We can compute this delay in the framework of this simple model if we make the first-order approximation that the average free-flow speed v is the same for everyone. The total delay due to congestion is then given by

$$\delta\tau = \frac{1}{v} \sum_{i,j} d_{ij} \left(\frac{T(j)}{c}\right)^{\mu}. \qquad (3.78)$$

If we assume that all the centers share the same number of commuters, a reasonable assumption within this model, we obtain

$$\delta\tau \sim \frac{L_{\text{tot}}}{v} \left(\frac{P}{h}\right)^{\mu} \qquad (3.79)$$

which, using the expressions for h, L_{tot} and A given in Eq. (3.72), (3.75), (3.76) respectively, gives

$$\delta\tau \sim \frac{\ell P}{v} \left(\frac{P}{c}\right)^{\delta} \qquad (3.80)$$

(in the fitting assumption Eq. (3.69), and using the same arguments for the calculation of $\delta\tau$, we easily obtain for the exponent the value $1+\frac{\mu}{\mu+1}(1-\frac{b}{2})$).

The total commuting time corresponding to the same distance but without congestion scales as $\tau_0 \sim L_{\text{tot}}$ and thus less rapidly than the total delay, which scales superlinearly with population (even when polycentricity is taken into account). This means that, as cities grow, delays due to congestion will eventually dominate the time spent in traffic, so that economic losses per capita due to the time lost in congestion increase with the size of the city.

3.4.5 CO_2 emission and gasoline consumption

Another diseconomy associated with congestion is the quantity of CO_2 emitted and gasoline consumed by motor vehicles. This amount depends not only on the

3.4 Revisiting the Fujita–Ogawa model

distance driven, but also on the traffic during the journey. It indeed turns out that for the same distance, a car burns more gas when the traffic is heavy than when it is clear. Within the model above, the presence of traffic is seen in the time spent to cover a given distance, and we assume that the quantity of CO_2 emitted by a vehicle is essentially proportional to the total time spent in traffic, leading to

$$Q_{CO_2} = q \sum_{i,j} d_{ij} \left[1 + \left(\frac{T(j)}{c} \right)^{\mu} \right] \qquad (3.81)$$

where q is the average quantity of CO_2 produced per unit time.

In the polycentric case with $h = h(P)$ subcenters, the typical trip length $\overline{d_{ij}}$ is given by $\sqrt{A/h}$ and we obtain

$$Q_{CO_2} = q \ell P \left[1 + \left(\frac{P}{c} \right)^{\delta} \right]. \qquad (3.82)$$

The quantity of CO_2 is thus dominated by congestion effects at large populations

$$Q_{CO_2} \sim q \ell P \left(\frac{P}{c} \right)^{\delta} \qquad (3.83)$$

and the total daily transport-related CO_2 emission per capita thus scales as

$$\frac{Q_{CO_2}}{P} \propto q \ell \left(\frac{P}{c} \right)^{\delta}. \qquad (3.84)$$

The quantity of CO_2 emitted per capita in cities thus increases with the size of the city, a consequence of congestion. This prediction agrees with the exponent we measure (see Fig. 3.10) on data gathered for United States and OECD cities.

The scaling of the CO_2 with population size is controversial, with results varying from one study to another (see also Section 2.5). Some studies (Fragkias et al. 2013) are concerned with the total emissions of CO_2, while here we consider emissions due to transport only. The prediction obtained here agrees well with the exponent of 1.33 measured by Oliveira et al. (2014) on the same dataset, but with a different definition of cities, and with measurements made by Rybski et al. (2013) for developing countries.

This superlinear increase of CO_2 emissions has terrible consequences on the economy, the environment, health, and well-being. If nothing is done and if the individual car stays the dominant transportation mode, cities will put even more strain on populations and participate in the overall increase of the planet's average temperature (Oreskes 2004). Advantages asssociated with urban living still outweigh the costs, but this superlinear increase shows that this will very

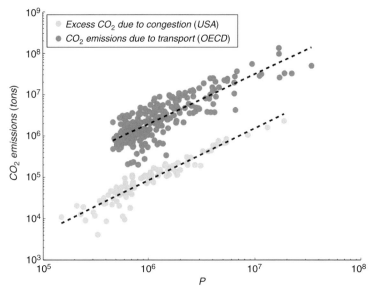

Figure 3.10. Variation of CO_2 emissions due to transport with city size. In light gray, excess CO_2 (in tons) due to congestion, as given by the Urban Mobility Report (2010) for 101 metropolitan areas in the United States. In dark gray, we show the estimated CO_2 emissions (in tons) due to transport, as given by the OECD for 268 metropolitan areas in 28 different countries. The dashed lines represent the power-law fit, which gives $Q_{CO_2} \sim P^{1.262 \pm 0.089}$ ($r^2 = 0.94$) for the United States data and $Q_{CO_2} \sim P^{1.212 \pm 0.098}$ ($r^2 = 0.83$) for the OECD data. Figure taken from Louf and Barthelemy (2014a).

rapidly cease to be the case, calling into question the sustainability of large cities. Mobility is at the heart of the problem and in any case it seems necessary to reduce the share of individual vehicles (being electrical or not), or to consider the development of smaller or medium-sized cities. In this case infrastructure costs are larger but the impact on the environment and on the well-being of individuals would be beneficial.

Another important related quantity is the the consumption of gasoline, which in principle is proportional to the emission of CO_2 and the time spent driving. The total daily gasoline consumption is then given by

$$Q_{\text{gas}} \sim q \, \ell \, P \left(\frac{P}{c} \right)^{\delta} \tag{3.85}$$

where q is the average quantity of gasoline needed per unit time. From this expression, the total daily gasoline consumption per capita scales as

$$\frac{Q_{\text{gas}}}{P} \sim \ell \sqrt{\frac{P}{\rho}} = \ell \sqrt{A} \tag{3.86}$$

and is therefore not a simple function of the city density, in contrast with what was suggested by the seminal paper of Newman and Kenworthy (1989). At this stage, however, more data about gasoline consumption is needed to test this prediction and draw definitive conclusions.

3.4.6 Urban villages

As some diseconomies associated with living in cities grow superlinearly, we may wonder if the economies of scale realized in larger cities can always counterbalance them. In order to have some estimates, we assume that the total cost of a city of population P is the sum of its infrastructure cost and the economic losses due to congestion

$$C_T(P) = \epsilon_I L_N(P) + \epsilon_C \Delta t \, \delta\tau(P) \tag{3.87}$$

where ϵ_I is the average cost of a kilometer of road, ϵ_C the average hourly wage and Δt the planning horizon in years (this expression is not exhaustive, as the costs dues to CO_2 emissions and gasoline consumption are not included). The infrastructure needs maintenance, and its cost depends on the planning horizon as well and can be written $\epsilon_I = \epsilon_B + \Delta t \, \epsilon_M$, where ϵ_B is the construction cost in \$/km and ϵ_M the maintenance cost in \$/km/year.

We compare this total cost with a system made of n cities of size P/n. The total traffic lane miles for the n cities reads $L_N^{(n)}(P) = n L_N(P/n)$, where $L_N(P) \sim \ell \sqrt{P} \left(\frac{P}{c}\right)^\delta$ is the total lane length for one city. The total congestion delay for n cities is $\delta\tau_n = n\delta\tau(P/n)$ and we thus obtain the total cost $C_T(P,n)$ for n cities

$$C_T(P,n) = n^{-\delta} \left[\ell \epsilon_I \sqrt{\frac{n}{P}} + \tau \epsilon_C \Delta t \right]. \tag{3.88}$$

The number of cities $n_{\min}$ that minimizes the total cost is obtained when $\frac{dC}{dn} = 0$, leading to (for $\Delta t \gg 1$)

$$n_{\min} = P \left[\left(\frac{2\delta}{1+\delta} \right) \frac{\epsilon_C}{\epsilon_M} \frac{\tau}{\ell} \right]^2 \tag{3.89}$$

(the actual number of cities is of course an integer, and can be taken as the nearest integer from $n_{\min}$, for instance). It is then economically advantageous to divide the population into several cities if $n_{\min} > 2$. To illustrate this point, we compute the number of cities which would minimize the cost for a world population $P \approx 10^9$. The World Bank estimates the maintenance cost of roads to be of the order of $\epsilon_M \approx 10^5$ \$/km/year, and the average hourly wage to be of the order of $\epsilon_C \approx 10$ \$/h; the value of δ is taken from the measures on US data, $\delta \approx 0.27$, and $\tau/\ell \approx 10$ km/h. We then obtain

$$n_{\min} \approx 180 \qquad (3.90)$$

which gives an average city size of $P/n \approx 5,500,000$. This result needs to be put in perspective with the fact that the world hosts 40 or so cities with over $5,500,000$ inhabitants and that this number is still increasing.

3.4.7 The most economical population distribution

The previous results assume that we split a large city into many cities of the same size. Cities are, however, organized in various sizes distributed according to a distribution described by the Zipf law (Zipf 1949). As we shall discuss in Chapter 8, it is still relatively unclear why we observe such a convergence to a "universal" law independent of cultural and historical specifics (Batty 2006; Cristelli et al. 2012). Although a dynamical aspect is certainly crucial in this question, we can consider it from the point of view of total cost optimization. More precisely, we assume that the populations of cities are distributed according to a power-law distribution

$$\mathcal{P}(x) = (\gamma - 1)x^{-\gamma} \quad \text{for } x \in [1, \Lambda] \qquad (3.91)$$

with $\gamma > 1$ and a cut-off population $\Lambda \gg 1$ (which is at most equal to the world's population). We can then ask what value of the exponent γ minimizes the overall cost, and from the discussion above the total cost for a population size x is given over large time scales by

$$C_T(x) = \epsilon_M \Delta t \, L_N(x) + \epsilon_C \, \Delta t \, \delta\tau(x). \qquad (3.92)$$

Averaging over the population distribution, we obtain

$$\overline{C_T} = \int_1^\Lambda \mathcal{P}(x) \, C_T(x) \, dx \qquad (3.93)$$

leading to

$$\overline{C_T} = \frac{\Delta t \, \ell}{c^\delta} (\gamma - 1) \left[\epsilon_M \frac{\Lambda^{-\gamma+\delta+\frac{3}{2}} - 1}{-\gamma + \delta + \frac{3}{2}} + \frac{\epsilon_C}{v} \frac{\Lambda^{-\gamma+\delta+2} - 1}{-\gamma + \delta + 2} \right]. \qquad (3.94)$$

The only consistent solution is obtained for $\gamma < \delta + 2$. The dominant term for $\Lambda \gg 1$ is given by

$$\overline{C_T} \simeq \frac{\Delta t \, \epsilon_C \, \ell}{c^\delta} (\gamma - 1) \left[\frac{\Lambda^{-\gamma+\delta+2}}{-\gamma + \delta + 2} \right]. \qquad (3.95)$$

3.4 Revisiting the Fujita–Ogawa model

The optimal power-law distribution minimizes the average cost and is such that $\frac{d\overline{C_T}}{d\gamma} = 0$, so we obtain the following equation

$$\frac{1+\delta}{\delta+2-\gamma} = (\gamma - 1)\ln \Lambda \tag{3.96}$$

and in the limit $\Lambda \gg 1$ the optimal value for γ is

$$\gamma^* = 2 + \delta - \frac{1}{\ln \Lambda}. \tag{3.97}$$

Numerically, $\delta \approx 0.32$ and $\Lambda \approx 10^9$, leading to $\gamma^* \approx 2.27$. It is interesting to note that this value would lead to a rank-plot exponent (≈ 0.78) not far from those measured on different countries around the world (Soo 2005). Although this argument does not pretend to provide a definitive answer to the Zipf puzzle, it nevertheless suggests that the broad diversity of population has an impact on the total cost and is not unconnected to economic considerations.

4
Infrastructure networks

Infrastructure such as transportation networks for individuals and freight, power grids or distribution systems is crucial for our societies and the good functioning of cities. All these networks are embedded in space: nodes have a position and links have a certain length, and hence a cost associated with their formation and maintenance. These "spatial networks" (Barthelemy 2011) necessitate specific tools for their characterization and in this chapter we first review some of the most important ones. We then focus on transportation networks, starting with the road and street network, followed by subway networks. We end this chapter with a digression on the railroad network and we discuss the importance of the spatial scale and the main differences between subways and railroads.

4.1 Roads and streets: patterns

An important component of cities is their street network (in the following we will not make the distinction between streets and roads). These networks are made of nodes that represent the intersections and the links are segments of roads between consecutive intersections. These networks can be thought as a simplified view of cities, that captures a large part of their structure and organization (Southworth and Ben-Joseph 2003), and contains a large amount of information about underlying and universal mechanisms at play in their formation and evolution. Identifying the main mechanisms in these systems is not a new task (Haggett and Chorley 1969; Xie and Levinson 2011), but the recent availability of digitized maps, historical or contemporary (see Chapter 1), allows us to test ideas and models on large-scale cross-sectional and historical data.

Street networks are approximatively planar graphs and are now fairly well characterized (Jiang and Claramunt 2004; Rosvall et al. 2005; Porta et al. 2006b,a; Lämmer et al. 2006; Crucitti et al. 2006; Cardillo et al. 2006; Xie and Levinson 2007; Jiang 2007; Masucci et al. 2009; Chan et al. 2011; Courtat et al. 2011). Most

4.1 Roads and streets: patterns

of their properties are essentially due to spatial constraints (Barthelemy 2011) and we can list the most common ones:

- the degree distribution is peaked (see for example Fig. 4.4(a)),
- the clustering coefficient and assortativity are large,
- so far, most of the interesting information lies in the spatial distribution of betweeenness centrality.

An important point is that information about these networks is not contained in their adjacency matrix only. Geometry, encoded in the spatial distribution of nodes, plays a crucial role. This implies in particular that a classification of cities according to their street network should rely on both topology and geometry. While classifications do not provide any understanding of the objects being classified, they provide a useful first insight about the different characteristics exhibited by objects of the same nature. Classifying, from a fundamental point of view, is however difficult: finding a typology of street patterns essentially amounts to classifying planar graphs, a non-trivial problem. The classification of street networks has been previously addressed by the space syntax community (Hillier and Hanson 1984; Penn 2003) and a good account can be found in the book by Marshall (2004). These works, although based on empirical observations, contain a large measure of subjectivity and it is important to eliminate this subjective part in order to reach a non-ambiguous, scientific classification of these patterns.

An interesting direction is provided by the study of leaves and their classification according to their venation patterns (Katifori and Magnasco 2012; Mileyko et al. 2012), but with a notable difference which prevents us from a direct application to streets: the existence of a hierarchy of veins governed by their diameter (the width of streets is usually absent from datasets). Another important direction could be provided by combinatorics: an exact bijection can be constructed between planar graphs and trees (Bouttier et al. 2004). Using this bijection, classifying planar graphs would amount to classifying trees, which is a simpler problem. However, the bijection does not take into account the geometrical shape of the planar graph: indeed, two street patterns can have the same topology but cells could be of very different areas, leading to visually different patterns and to cities of different spatial structure. It is thus important to take into account not only the topology of the planar graph – as described by the adjacency matrix – but also the position of the nodes.

4.1.1 Length of the network

The total length of the network given by $\ell_T = \sum_e \ell(e)$ (where $\ell(e)$ is a measure of the length of the edge e) and is an interesting measure, as it contains some information about the level of development of the city. In Fig. 4.1, the total length

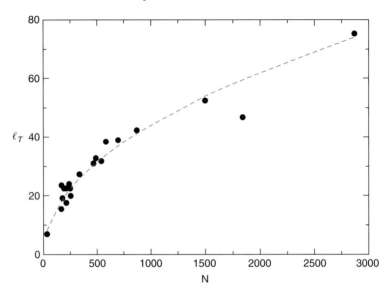

Figure 4.1. Total length versus the number of nodes. The line is a fit which predicts a growth as $\sqrt{N}$. Figure from Barthelemy and Flammini (2008).

ℓ_T of the network versus the number of nodes N for a set of cities (Cardillo et al. 2006) can be fitted by a power function of the form

$$\ell_T = \mu N^\beta \qquad (4.1)$$

with $\mu \approx 1.51$ and $\beta \approx 0.49$. In order to understand this result, one has to focus on the street segment length distribution $P(\ell_1)$. This quantity has been measured for London in Masucci et al. (2009) and is shown in Fig. 4.2. This figure shows that the distribution decreases rapidly and the fit proposed by Masucci et al. suggests that

$$P(\ell_1) \sim \ell_1^{-\gamma} \qquad (4.2)$$

with $\gamma \simeq 3.36$, which implies that both the average and the dispersion are well-defined and finite. If we assume that this result extends to other cities, it means that we have a typical distance ℓ_1 between nodes which is meaningful. This typical distance between connected nodes then naturally scales as

$$\ell_1 \sim \frac{1}{\sqrt{\rho}} \qquad (4.3)$$

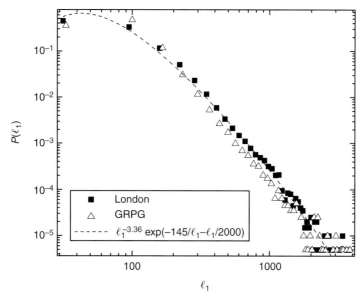

Figure 4.2. Length distribution $P(\ell_1)$ for the street network of London (and for the model GRPG proposed in this study). From Masucci et al. (2009)).

where $\rho = N/L^2$ is the density of vertices and L the linear dimension of the ambient space. This implies that the total length scales as

$$\ell_T \sim E\ell_1 \sim \frac{\langle k \rangle}{2} L\sqrt{N} \qquad (4.4)$$

which is in good agreement with the $\sqrt{N}$ behavior observed in Fig. 4.1 and also for the value of the prefactor $\mu \approx \langle k \rangle/2$.

4.1.2 Statistics of blocks

When we are looking at a street map of a city, what is essentially perceived by the eye is not deriving from streets but from the distribution of the shape, area and orientation of blocks. Blocks are defined (see Fig. 4.3) as the cells of the planar graph formed by streets, and are easy to extract from a map. In order to compare the street patterns of different cities, it thus seems natural to consider the distribution of blocks and to construct a classification based on this quantity, rather than streets. A block can be defined without ambiguity as being the smallest area delimited by roads (it has then to be distinguished from a parcel, which is a tax-related definition). While the information contained in the blocks and the streets are equivalent (up to dead-ends), the information related to the visual aspect of the street network seems to be easier to extract from blocks. Blocks are indeed simple geometrical objects – polygons – whose properties are easily characterized

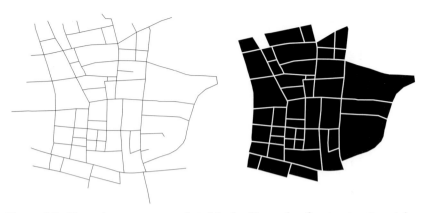

Figure 4.3. From the street network to blocks. Example of a street pattern taken in the neighborhood of Shibuya in Tokyo (Japan) and the corresponding set of blocks (note that the block representation does not take into account dead-ends). Figure from Louf and Barthelemy (2014a).

by quantities such as their area and shape. In the following we consider these different properties and show that an accurate description of street patterns can be obtained with a combination of the area and the shape.

Area distribution

The surface area A of a block gives a useful indication, and its distribution is an important piece of information about the block pattern. As shown by Lämmer et al. (2006) and Fialkowski and Bitner (2008), different cities have blocks with an area distribution that decreases as a power law

$$P(A) \sim \frac{1}{A^\tau} \tag{4.5}$$

where the exponent is of order $\tau \approx 2$ (Lämmer et al. 2006; Barthelemy 2011; Strano et al. 2012; Barthelemy et al. 2013). On the Fig. 4.4(b), we show this distribution obtained for the city of Dresden (Germany) obtained by Lämmer et al. (2006). Although this seemingly universal behavior gives a useful constraint on any model that attempts to model the evolution of cities' road networks, it does not allow cities to be distinguished from each other. This result is in sharp contrast to the simple picture of an almost regular lattice, for which the distribution $P(A)$ is peaked around ℓ_1^2. It is interesting to note that if we assume that $A \sim \ell_1^2 \sim 1/\rho$, where ρ is the density of nodes and is distributed according to a distribution $f(\rho)$ (with a finite $f(0)$), a simple calculation gives

$$P(A) \sim \frac{1}{A^2} f(1/A) \tag{4.6}$$

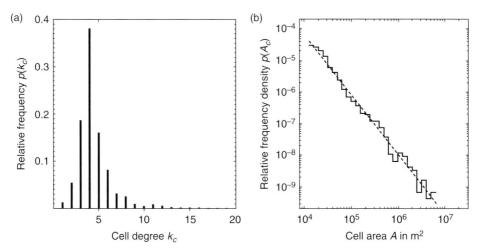

Figure 4.4. (a) Degree distribution. (b) The frequency distribution of the cells with surface areas A obeys a power law with exponent $\tau \approx 1.9$ (for the road network of Dresden). Figure from Lämmer et al. (2006).

which behaves as $P(A) \sim 1/A^2$ for large A. This simple argument thus suggests that the observed value $\tau \approx 2.0$ of the exponent is universal and reflects the fluctuations of the local density. In the following, we will show how a simple fragmentation model can also explain this exponent of order 2.

The shape of blocks

A second characterization of a block is through its shape, with the form (or shape) factor Φ, defined in the quantitative geography literature (Haggett et al. 1977) as the ratio between the area A of the block and the area A_C of the circumscribed circle C

$$\Phi = \frac{A}{A_C} \tag{4.7}$$

(the denominator can also be taken as $\pi D^2/4$, where D is the largest length in the polygon). The quantity Φ is always smaller than one, and the smaller its value, the more anisotropic the block is. There is not a unique correspondence between a particular shape and a value of Φ, but this measure gives a good indication about the block's shape in real-world data, where most blocks are relatively simple polygons. For rectangles with sides a and b, isosceles triangles with sides a and b

$(b > a)$, we easily obtain

$$\Phi_\square = \frac{4}{\pi} \frac{ab}{a^2 + b^2}$$

$$\Phi_\square = \frac{2}{\pi} \approx 0.64$$

$$\Phi_\triangle = \frac{4}{\pi} \frac{a}{b}$$

$$\Phi_\bigcirc = 1.$$

These values show that typically for $\Phi > 0.5$ we have rectangles and squares, and for values below that we have essentially triangles and other elongated shapes. Lämmer et al. (2006) found for German cities that most of the blocks have a form factor between 0.3 and 0.6, suggesting a large variety of cell shapes.

The fingerprint of cities

The distribution of Φ displays important differences from one city to another, and a first, naïve idea would be to classify cities according to the distribution of block shapes given by $P(\Phi)$. The shape itself is however not enough to account for visual similarities and dissimilarities between street patterns. Indeed, for cities such as New York and Tokyo for example, even if we observe similar distributions $P(\Phi)$ (see Fig. 4.5), the visual similarity between the two cities' layouts is not obvious at all. A reason for this is that blocks can have a similar shape but very different areas: if two cities have blocks of the same shape in the same proportion, but with totally different areas, they will look different. We thus need to combine the information about both the shape and the area.

In order to construct a simple representation of cities which integrates both area and shape, we rearrange the blocks according to their area (on the y-axis) and display their Φ value on the x-axis (Fig. 4.5). We divide the range of areas in to logarithmic bins and the gray level of a block represents the area category to which it belongs. We plot the conditional probability distribution $P(\Phi|A)$ of shapes, given an area bin (Fig. 4.5, right) where the curves represent the distribution of Φ in each area category, and where the gray area is the sum of all the these curves and is the distribution of Φ for all cells. This is simply the translation of the well-known formula for probability conditional distribution

$$P(\Phi) = \sum_A P(\Phi|A) P(A).$$

These figures give a "fingerprint" of the city which encodes information about both the shape and the area of the blocks. In order to quantify the distribution of

4.1 Roads and streets: patterns

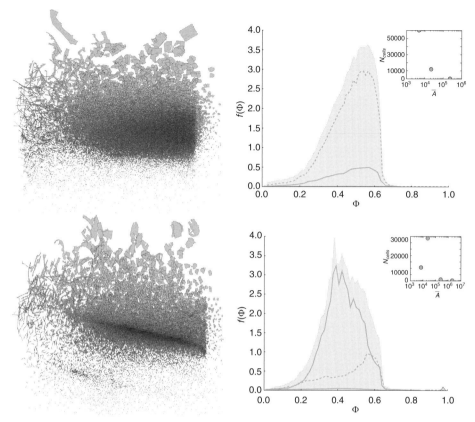

Figure 4.5. The fingerprints of Tokyo (top) and New-York, NY (bottom). (Left) We rearrange the blocks of a city according to their area (y-axis), and their Φ value (x-axis). The gray level of each block corresponds to the area category it falls into. (Right) We quantify this pattern by plotting the distribution of shapes, as measured by Φ for each area category, represented by the different continuous and dashed curves. The gray surface is the sum of all the curves and represents the distribution of Φ for all cells. The continuous line represents medium size blocks and the dotted line small blocks. Figure from Louf and Barthelemy (2014c).

blocks inside a city, and thus the visual aspect of the latter, we then use $P(\Phi|A)$ for different area bins. The comparison between these quantities provides the basis for a classification of street patterns.

If we consider that two cities display similar patterns if their blocks have both similar area and shape, the shape distributions for each area bin should be very close, and this simple idea allows us to propose a distance between street patterns of different cities. We sort the blocks according to their area in distinct bins (two in the case considered in Louf and Barthelemy 2014c, the small and medium-sized blocks) and we denote by $f_\alpha(\Phi)$ the ratio of the number of cells with a form factor

Φ that lie in the bin α over the total number of cells for that city. The distance d_α between two cities a and b characterized by their respective function f_α^a and f_α^b is then

$$d_\alpha(a,b) = \int_0^1 \left| f_\alpha^a(\Phi) - f_\alpha^b(\Phi) \right|^n d\Phi \tag{4.8}$$

and by combining all area bins α, we can then construct a distance D between two cities given by

$$D(a,b) = \sum_\alpha d_\alpha(a,b)^2. \tag{4.9}$$

At this point, we have a distance between the patterns of two cities and we can perform a hierarchical clustering (see for example Kaufman and Rousseeuw 1990) on many cities. At an intermediate level, four groups of cities emerge that are easily interpretable in terms of the abundance of blocks of a given shape and area. In Fig. 4.6 we show the average distribution of Φ for each category and show the typical street pattern associated with each of these groups. The main features of each group are the following:

- In group 1 (comprising Buenos Aires, Argentina only) we essentially have blocks of medium size with shapes that are dominated by the square shape and regular rectangles. Small area blocks are almost exclusively squares.
- Athens (Greece) is a representative element of group 2, which comprises cities with a dominant fraction of small blocks with shapes broadly distributed.
- Group 3 (illustrated here by New Orleans, USA) is similar to group 2 in terms of the diversity of shapes but is more balanced in terms of areas, with a slight predominance of medium-sized blocks.
- Group 4, which contains for this dataset the interesting example of Mogadishu (Somalia), displays essentially small, square-shaped blocks, together with a small fraction of small rectangles.

There are certainly important sampling effects here and it would be worth redoing this classification with more cities, but we can already try to understand some aspects of these different groups. For example, all North American cities (except Vancouver, Canada) are part of group 3, as well as all European cities (except Athens, Greece). The composition of the other continents is more balanced between the different groups. At a smaller scale within group 3, all European cities (but Athens) considered here belong to the same subgroup of group 3. Similarly, 15 American cities out of the 22 in the dataset belong to the same subgroup of group 3. Exceptions are Indianapolis (IN), Portland (OR), Pittsburgh (PA), Cincinnati (OH), Baltimore (MD), Washington (DC), and Boston (MA), which are classified with European cities, confirming the impression that these US cities have a European

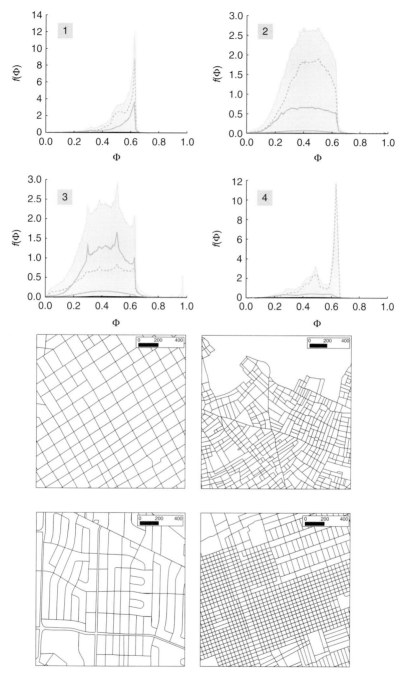

Figure 4.6. Four groups of patterns. (Top) Distribution of the shape factor Φ for each group found by the clustering algorithm. The continous curve gives the shape distribution for medium-sized blocks and the dashed line for small blocks. (Bottom) Typical street pattern for each group (plotted at the same scale in order to observe differences both in shape and areas). Group 1 (top left): Buenos Aires — Group 2 (top right): Athens — Group 3 (bottom left): New Orleans — Group 4 (bottom right): Mogadishu. Figure from Louf and Barthelemy (2014c).

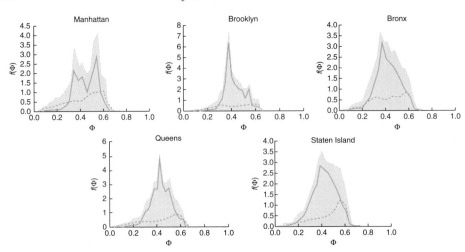

Figure 4.7. Fingerprints for the borough of New York (the continuous curve gives the shape distribution for medium-sized blocks and the dashed line for small blocks). Only Staten Island and the Bronx have similar fingerprints, while the others are different. In particular, Manhattan exhibits two sharp peaks at $\Phi \approx 0.3$ and $\Phi \approx 0.5$, which are the signature of a grid-like pattern with the predominance of two types of rectangle. Brooklyn and Queens exhibit a sharp peak at different values of Φ, signaling the presence of grid-like patterns made of different basic rectangles. Figure taken from Louf and Barthelemy (2014c).

feel. These results point towards important differences between US and European cities, and could constitute the starting point for the quantitative characterization of these differences (Bretagnolle et al. 2010).

The classification of patterns presented here represents a step towards a quantitative and systematic comparison of cities, but it is certainly not the end of the story. Indeed, cities are usually made of different neighborhoods which often exhibit different street patterns. In Europe, the division is usually clear between the historical center and the more recent suburban areas of the city. The fingerprint of a large city is then the combination of different fingerprints corresponding to smaller coherent neighborhoods. In the case of New-York, while Staten Island and the Bronx have very similar fingerprints, the others are different (see Fig. 4.7). Manhattan exhibits two sharp peaks at $\Phi \approx 0.3$ and $\Phi \approx 0.5$, which are the signature of a grid-like pattern with the predominance of two types of rectangle. Brooklyn and the Queens exhibit a sharp peak at different values of Φ, also the signature of grid-like patterns with different rectangles for basic shapes. A further step in the classification would thus be to find a method to extract these neighborhoods, and integrate the spatial correlations between different types of neighborhood.

More importantly, these results should be combined with the specific knowledge of architects, and urbanists in order to identify the main factors that govern the statistics of street patterns. Indeed, further studies are certainly needed in order to relate the various types that we observe to different urban processes. For example, in some cases, small blocks are obtained through a fragmentation process and their abundance could be related to the age of the city. A large regularity of cell shapes could be related to planning, such as in the case of Manhattan for example, but we also know from the example of Paris (Barthelemy et al. 2013) that a large variety of shapes is also directly related to the effect of an urban modification that does not respect the existing geometry.

4.1.3 Structure of paths

Betweenness centrality

The betweenness centrality (BC) introduced by Freeman (1977) is one of the important quantities in studying complex networks (Newman 2001; Barthelemy 2004) and for street networks in particular (Crucitti et al. 2006). It quantifies the importance of a node (or a link) by the amount of traffic going through it, assuming uniform demand where the traffic between all pairs of nodes is the same. More precisely, given a graph $G \equiv G(V, E)$, the betweenness centrality $g(e)$ of the link e is defined as

$$g(e) = \sum_{s \in V} \sum_{t \in V, t \neq s} \frac{\sigma_{st}(e)}{\sigma_{st}} \qquad (4.10)$$

where σ_{st} is the number of shortest paths from node s to node t, while $\sigma_{st}(e)$ is the number of such shortest paths which contain the link e. This quantity measures how many times a link is used in the shortest paths connecting two different nodes of the network, and is thus a measure of the contribution of a link to the flow organization in the network. Note that with this definition, the betweenness centrality of terminal nodes is zero. The betweenness centrality can similarly be defined for nodes

$$g(i) = \sum_{s,t} \frac{\sigma_{st}(i)}{\sigma_{st}} \qquad (4.11)$$

where $\sigma_{st}(i)$ is the number of shortest paths going from s to t and going through the node i.

The BC is an especially relevant quantity for urban systems: in particular, it is correlated with income (Venerandi et al. 2014), the locations of shops and other micro-economic activity (Porta et al. 2009), and urban growth (Strano et al. 2012; Barthelemy et al. 2013). In the case of car traffic and congestion, the absence of detailed traffic models or mobility data can lead to use of the BC in order to identify the *potentially* congested locations and the effects of spatial structure on

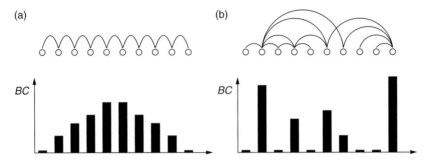

Figure 4.8. (a) Betweenness centrality for the (one-dimensional) lattice case. The central nodes are close to the barycenter. (b) For a general graph, the central nodes are usually the ones with large degree.

the shortest path structure. Even if the assumptions used in the BC calculation can lead to some inaccuracies (Wang et al. 2012), it is the simplest proxy that contains some level of information about real traffic. Also, the spatial distribution of the BC gives important information about the coupling between space and the topology of the road network.

In general, the BC of a vertex is determined by its ability to provide a path between separated regions of the network. Hubs in complex networks are natural crossroads for paths and it is natural to observe a marked correlation between the average over nodes with the same degree

$$g(k) = \frac{1}{N(k)} \sum_{i/k_i=k} g(i) \qquad (4.12)$$

(where $N(k)$ is the numbers of nodes with degree k) and k as expressed in the following relation (Barthelemy 2004):

$$g(k) \sim k^{\eta} \qquad (4.13)$$

where η depends on the characteristics of the network. We expect this relation to be altered when spatial constraints become important and in order to illustrate this effect we consider a one-dimensional lattice. For this lattice, the shortest path between two nodes is simply the euclidean geodesic and for two points lying far from each other, the probability that the shortest path passes near the barycenter of the network is very large. In other words, the barycenter (and its neighbors) have a large centrality, as illustrated in Fig. 4.8a. More precisely, if the lattice is defined as $x \in [0, L]$, the BC is simply given by $g(x) \sim x(L - x)$, which has indeed a maximum at $x = L/2$. In contrast, in a purely topological network with no underlying geography, this consideration does not apply anymore and if we rewire more and more links (as illustrated in Fig. 4.8b) we observe a progressive

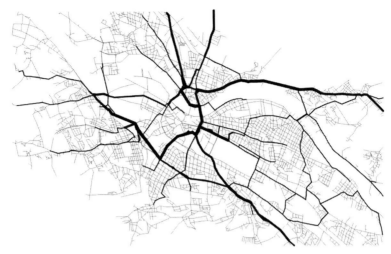

Figure 4.9. Betweenness centrality for the city of Dresden. The width of the links corresponds to the betweenness centrality. Figure from Lämmer et al. (2006).

decorrelation of centrality and space while the correlation with degree increases. In a lattice, the betweenness centrality depends essentially on space and is maximum at the barycenter, while in a network the betweenness centrality of a node depends on its degree. When the network is constituted of long links superimposed on a lattice, we expect the appearance of "anomalies", characterized by large deviations around the behavior $g \sim k^\eta$.

Lämmer et al. (2006) studied the German road network and obtained very broad distributions of betweenness centrality with a power-law exponent in the range [1.28, 1.49] (for Dresden $\approx$ 1.36). These broad distributions of the betweenness centrality signal the strong heterogeneity of the network in terms of traffic, with the existence of a few very central roads which most probably points to important traffic congestion problems. Also the absence of a scale in a power-law distribution suggests that the importance of roads is organized in a hierarchical way, a property expected for many transportation networks (Yerra and Levinson 2005). The broadness of the betweenness centrality distribution does not seem however to be universal. Indeed, Crucitti et al. (2006) and Cardillo et al. (2006) found that the betweenness centrality distribution is peaked (depending on the city either exponentially or according to a Gaussian), which signals the existence of a scale and therefore of a finite number of congested points in the city.

As mentioned earlier, the spatial distribution of the betweenness centrality is in itself interesting, since it points to the important zones which potentially are congested. Fig. 4.9 displays the spatial distribution of the betweenness centrality for the city of Dresden. As expected, zones which are central from a geographical point of view also have a large betweenness centrality. We see, however, that

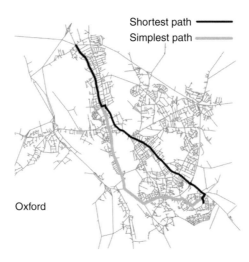

Figure 4.10. Example of shortest (black line) and simplest (gray line) paths illustrated on the Oxford (UK) street network. The simplest path has fewer turns at the expense of being longer than the shortest path. Figure taken from Viana et al. (2013).

other roads or zones can have a large betweenness centrality pointing to a complex pattern of flow distribution in cities. Further studies are needed in order to understand how the interplay between space and the topology leads to these patterns.

Simplicity

The BC is based on the structure of shortest paths, but we can actually define different types of path for a given pair of nodes (i, j). The usual quantity is the shortest euclidean path of length $\ell(i, j)$ which minimizes the total distance travelled to go from i to j, but we could ask for another type of path that minimizes for example the number of turns. These "simplest" paths are of length $\ell^*(i, j)$ and if there is more than one such path we choose the shortest one. Fig. 4.10 displays an example of the shortest and simplest path for a given pair of nodes on the Oxford, UK street network.

To identify the simplest path, we first convert the graph from the primal to the dual representation where each node corresponds to a straight line on the primal graph. These straight lines are determined by a continuity negotiation-like algorithm (Porta et al. 2006a). Edges in dual space, in turn, represent the intersection of straight lines on the primal graph (see Fig. 4.10).

The shortest paths are not always simple, and the statistical comparison of the lengths of the shortest $\ell(i, j)$ and simplest $\ell^*(i, j)$ paths provides non-trivial and non-local information about the spatial organization of these graphs. It is then

4.1 Roads and streets: patterns

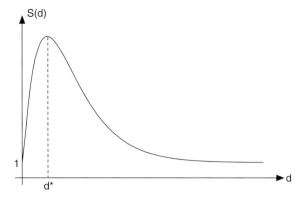

Figure 4.11. Typical shape of the simplicity profile with at least one maximum for a value d^*.

natural to introduce the *simplicity index* S as the average:

$$S = \frac{1}{N(N-1)} \sum_{i \neq j} \frac{\ell^*(i,j)}{\ell(i,j)}. \qquad (4.14)$$

The simplicity index is by definition larger than one and exactly equal to one for a regular square lattice and any tree-like network, for example. Large values of S indicate that the simplest paths are on average much longer than the shortest ones, and that the network is not easily navigable (we do not take into account here congestion effects, which can influence the path choice). This metric is a first indication of the spatial structure of simplest paths but mixes various scales, and in order to obtain more detailed information, we define the *simplicity profile*:

$$S(d) = \frac{1}{N(d)} \sum_{i,j/d_E(i,j)=d} \frac{\ell^*(i,j)}{\ell(i,j)}, \qquad (4.15)$$

where $d_E(i,j)$ is the euclidean distance between i and j and where $N(d)$ is the number of pairs of nodes at euclidean distance d. This quantity $S(d)$ is also larger than one and its variation with d informs us about the large-scale structure of these graphs. We can draw a generic shape of this profile (Fig. 4.11). For small d, at the scale of nearest neighbors, there is a large probability that the simplest and shortest paths have the same length, yielding $S(d \to 0) \sim 1$, and increasing for small d. For very large d, it is almost always beneficial to take long straight lines when they exist, thus reducing the difference between the simplest and the shortest paths. As a result we expect $S(d)$ to decrease when $d \to d_{max}$ (note that a similar behavior is observed for another quantity, the route-length efficiency, introduced by Aldous et al. 2010). The simplicity profile then displays in general at least one maximum at an intermediate scale d^* for which the length differences between the shortest and the simplest path is maximum. The length d^* thus represents the typical size of domains not crossed by long straight lines. At this intermediate scale, the detour needed to find long straight lines for the simplest paths is very large.

In Fig. 4.12 we show the simplicity profile for street networks of cities (Bologna, Italy; Oxford, UK; Nantes, France), the national highway network of Australia, the UK national railroad system, and the water supply network of central Nantes (France). We also show $S(d)$ for some biological networks such as the venation patterns of leaves (*Ilex aquifolium* and *Hymenanthera chatamica*), and of a dragongly wing (see Viana et al. 2013 for details). We observe that for most of these systems, the simplicity profile displays the generic shape with a maximum at an intermediate scale. In urban cases, such as Bologna and central Nantes, we have a typical monocentric system with a dense center and a few important radial straight lines, leading to a simple profile $S(d)$. In the case of Oxford and the Australian highway network, the polycentric organization leads to multiple peaks in the simplicity profile. Interestingly, we observe that the profiles for Australian highways and for railroads in the UK are very different, despite their similar scale. In particular, the UK railroad displays small values of the simplicity (less than ≈ 1.2), while for the Australian highway network there are many pairs of nodes for which the simplest path is much longer than the shortest one. We also observe that the profiles for both street and water systems of Nantes have a very similar shape, pointing to the fact that these networks are strongly correlated. In addition, the position and the height of the peak (≈ 1.4) observed for the Nantes water system suggests that this distribution system has similar features when compared to biological systems such as vein networks in leaves, whose function is also distribution.

Compared to urban systems, the simplicity profiles of biological networks have a single well-defined, and much more pronounced, peak. We observe values of order $S_{max} \approx 1.5$ and 2.5 for $d^*/d_{max} \approx 0.2$, meaning that for this range of distance, the detour made by the simplest path is very large. This peak is related to the existence of domains of typical size d^* not crossed by large veins. We see here a clear effect of the existence of the spatial organization of long straight lines in these systems, probably optimized for the distribution (of water for leaves). The decay for large d is also much faster in the biological case compared to urban systems: this shows that in biological systems there are long straight lines allowing the connection of far-away nodes. This is particularly evident on the leaves shown in Fig. 4.12, where we can see the first-levels veins (primary and secondary), the rest forming a network. For streets, the organization is much less rigid and the hierarchy less strict: we have a more uniform spatial distribution of straight lines, leading to a smoother decrease of $S(d)$.

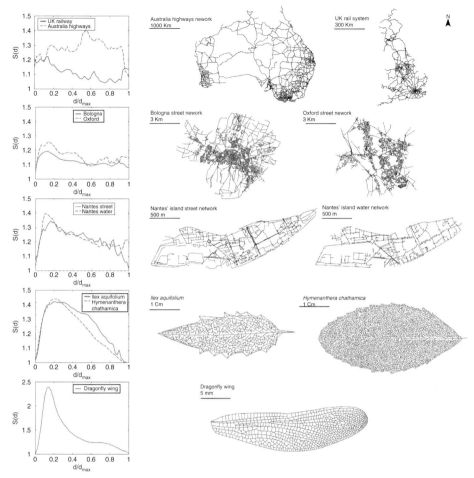

Figure 4.12. We represent here the simplicity profiles for different networks ranging from large-scale networks (10^6 m) to small scales of order $10^{-3}m$. We see in these different examples the effect of the presence of long straight lines and of a polycentric structure. In particular for the biological cases (leaves and dragonfly wing), we can clearly see that the peak at $d^* \sim 0.2 d_{max}$ corresponds to the size of domains not crossed by long straight lines. Figure taken from Viana et al. (2013).

4.2 Evolution of the road network

So far, we have discussed properties on static networks, and here we characterize the evolution of the road network. The first example of a time-evolving network is the road network of the Groane region, which is a 125 km^2 area located north of Milan (Strano et al. 2012) and for which there are 7 snapshots for different times from 1833 to 2007. This region evolved without central planning and is thus a good example of an "organic" evolution of urban systems.

The second example at a smaller scale is the evolution of central Paris between 1789 and 1999. This dataset provides an interesting case study, as Paris experienced large changes due to Haussmann in the middle of the nineteenth century (see Barthelemy et al. 2013 for details and more references about this network). This is an opportunity to observe quantitatively the effect of top-down planning: until 1836, we are in pre-Haussmann Paris, while from 1888 until now we are in the post-Haussmann period.

4.2.1 Basic properties

We first describe briefly the evolution of basic properties of the road network such as the number of nodes, of links, the total length of the network, the area and the shape factor of blocks.

In most cases during the evolution of the network, measured for example by the number of nodes N, the growth rate is not constant. For Groane, the total number N grows by a factor of twenty, from the original 255 nodes present at $t = 1833$ to more than $5,000$ nodes at $t = 2007$, with a period of slow growth from 1833 to 1933, fast from 1933 to 1994, and slow again from 1994 to 2007. Interestingly enough, the number of nodes N in the network is a linear function of the number of people living in the area or, in other words, the number of people per road intersection remains constant over time (Fig. 4.13a).

In the case of Paris, the number of nodes increased from about $3,000$ in 1836 to about $6,000$ in 1888 and the total length increased from about 400 kms to almost 700 kms, all this in about 50 years. The same behavior of N versus the population is also found in thid case: the number of nodes N is proportional to the population P, and the corresponding increase rate is of order $dN/dP \approx 0.0021$, similar to what was measured for the Groane area. This seems to reflect that the density of the street network indeed represents a good indicator of the evolution of a region and its economic potential, but we do not have a clear theoretical argument for this fact.

In order to understand the evolution of the network and to compare different cases, it thus seems natural to consider the number of nodes N as the natural internal clock of the system, and to study the change of various network properties as a function of N. In this way we get rid of exogenous effects on the evolution of this system.

We observe simple facts such as the number of links K, which grows almost linearly with N (Fig. 4.13b, top), showing that the average degree is roughly constant despite massive historical changes, with a slight increase from $\langle k \rangle \simeq 2.57$ to 2.8 when going from 1914 to 1980 (Fig. 4.13b, bottom). In the Paris case we

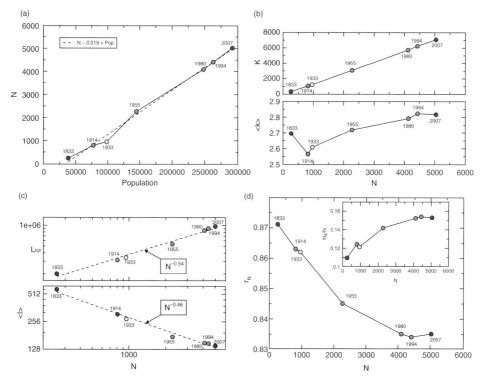

Figure 4.13. Groane area evolution. (a) Number of nodes N versus total population (continuous line with circles) and its linear fit (dashed line). (b) The total number of edges K, and the average node degree $\langle k \rangle$ as a function of N. The number of edges increases slightly faster than linearly with the number of nodes, as shown by the slight increase in the average node degree. (c) Both the total network length L_{tot} (upper panel) and the average link length $\langle l \rangle$ (bottom panel) scale as power-law functions of N (the corresponding fits are reported as dashed lines). (d) As the network grows, the value of the ratio r_N between the number of nodes with degree $k = 1$ and $k = 3$, and the total number of nodes, decreases, indicating the presence of a higher number of four-way crossings. In the inset we report the percentage of nodes having degree $k = 4$ as a function of N. Note that the relative abundance of four-ways crossings increases by 5% in two centuries. Figure from Strano et al. (2012).

also observe that K is proportional to N, giving an average degree of order 3 (Barthelemy et al. 2013).

The total network length (Fig. 4.13c) displays an increase of the form N^γ where $\gamma \simeq 0.54$. This result – also observed for the Paris case – is consistent with the evolution of two-dimensional lattices with a peaked link length distribution (Barthelemy and Flammini 2008; Barthelemy 2011), which are described by a value $\gamma = 1/2$. The quantity $r_N = [N(1) + (N(3)]/N$, where $N(k)$ is the number of nodes of degree k, gives additional information on the structure of the network.

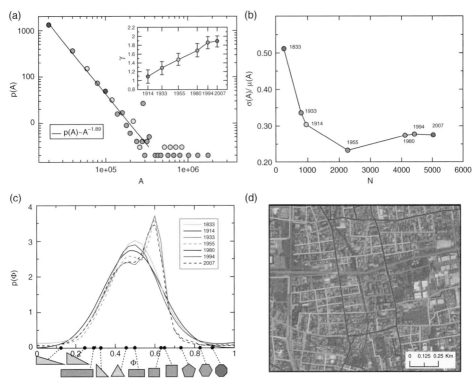

Figure 4.14. Groane area evolution. (a) The size distribution of cell areas at $t = 2007$ can be fitted with a power-law $p(A) \sim A^{-\gamma}$, with an exponent $\gamma \simeq 1.9$. The values of γ increase over the years as shown in the inset. (b) Relative dispersion in the distribution of areas as a function of the network size N. (c) Distribution of cell shapes at each time, as quantified by the shape factor Φ. We also report, as a reference, the values of shape factors corresponding to various convex regular polygons. (d) The map shows some typical cell shapes at different times, with the same color-code as in the previous panels. Figure from Strano et al. (2012).

Indeed, the plot of r_N versus N (Fig. 4.13d) shows that it steadily decreases from $r_N \simeq 0.87$ at $t = 1833$ to $r_N \simeq 0.835$ at $t = 2007$, signaling an evolution from a pre-urban to an urban phase.

In agreement with other results (Lämmer et al. 2006), the distribution of block areas at $t = 2007$ is a power law with the same exponent of order $\gamma = 2.0$ (see Fig. 4.14a). As reported in the inset, the exponent however changes in time: it takes a value $\gamma \simeq 1.2$ at $t = 1833$ and converges towards $\gamma \simeq 1.9$ as the network grows. Because a larger exponent indicates a higher homogeneity of cell areas, we are thus witnessing here a process of homogenization of the size of cells. This appears to be a clear effect of increasing urbanization in time, with the fragmentation of larger cells of natural land into smaller urbanized ones. Accordingly, the relative dispersion of cell areas, shown in Fig. 4.14b, decreases from 0.5 at $t = 1833$ to

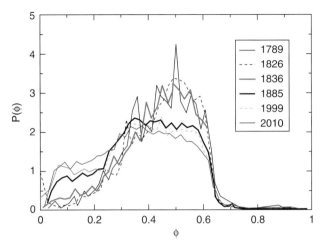

Figure 4.15. Shape factor distribution for central Paris at different times. Figure from Barthelemy et al. (2013).

0.26 at $t = 2007$, indicating that the variance of the distribution becomes smaller as N increases.

The distribution of the shape factor $P(\Phi)$ reported in Fig. 4.14c clearly reveals the existence of two different regimes: for $t \leq 1933$ the distribution is well approximated by a single Gaussian function with an average of about 0.5 and a standard deviation of 0.25. Conversely, for $t \geq 1955$ the distribution of shape factors displays two peaks and can be fitted by the sum of two Gaussian functions. The first peak coincides roughly with the one obtained for $t \leq 1933$, while the second peak, centered at 0.62, signals the appearance for $t \geq 1955$ of an important fraction of regular shapes such as rectangles with sides of similar lengths. In Fig. 4.14d some examples of the cell shapes at different times are shown. In the case of Paris we observe different distributions of $P(\Phi)$ according to the period considered (see Fig. 4.15). Until 1836, the distribution is stable and then we observe a dramatic change during the Haussmann period with a larger abundance of blocks with small values of Φ. These small values correspond to elongated rectangles or triangles created by streets crossing the existing geometry at various angles, and highlight the effect of central planning.

4.2.2 Simplicity profile

The simplicity profile shown in Fig. 4.16(a) allows us to distinguish two different periods for the Groane region. The first period from 1833 to 1955 displays a relatively small simplicity at all scales, while a distinct second regime appears from 1980 until now. In this latter regime, the simplicity profile is substantially

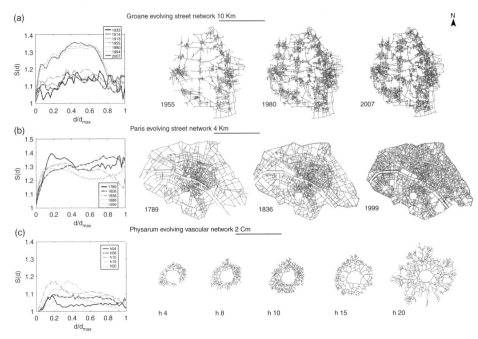

Figure 4.16. Simplicity profiles for time-varying networks. We represent here the profiles for (a) the road network of the Groane region (Italy), (b) the street network of Paris (France) in the pre-Haussmannian (1789, 1836) and post-Haussmannian (1999) periods, and in (c) the *Physarum* network growing on a period of one day approximately. We observe on (a) and (b) that the evolution of the profile is able to reveal important structural changes. Figure from Viana et al. (2013).

larger for all scales. This is an effect of the massive urban densification, leading to a polycentric structure where the readability and ease of navigation are drastically lowered.

The effect of Haussmann's central planning is clearly visible on the network shown in Fig. 4.16(b). From 1789 to 1836, we have a relatively large simplicity at all scales and we observe a decrease in that period at small scales ($d/d_{max} < 0.4$), which corresponds well to the fact that many religious and aristocratic domains and properties were sold and divided in order to create new houses and new roads, improving the navigation around Paris. The 1826–1836 transition displays a decrease of simplicity for distances larger than roughly 5 kms (corresponding to $d/d_{max} \approx 0.6$), indicating that long-distance routes were simplified. It is interesting to note that during this period the eastern part of Paris experienced large transformations with the construction of the St. Martin channel. Finally, in the period 1836 to 1888, when Paris experienced Haussmann's transformation, the simplicity profile is strongly affected: compared to 1836, the simplicity is improved in the range $d/d_{max} \in [0.3, 0.8]$, which can be attributed to the construction of large avenues connecting important nodes of the city. In addition, we observe the surprising effect that at large scales $d/d_{max} > 0.8$, the simplicity is degraded by Haussmann's work: this however could be an artifact of the method and the fact that we considered a portion of Paris only and neglected the effect of surroundings.

We note that differences between Groane and Paris might be explained in terms of a sparse, polycentric urban settlement (Groane) versus a dense one (Paris). In particular, in the "urban" phase for Groane (after 1955), the simplicity profile becomes similar to the one of a dense urban area such as Paris.

Finally, we can compare these results with a biological network and we show the evolution in Fig. 4.16(c) for the *Physarum Policephalum*, a biological system evolving at the centimeter scale. The simplicity profile for the Physarum is relatively low (less than ≈ 1.2), suggesting that simplicity could be an important factor in the evolution of this organism. A closer observation shows that during its evolution, the Physarum adds new links to the previous network and also modifies the network on a larger scale, as revealed by the changes of the simplicity profile. The evolution of the profile is similar to the one obtained for a null model (Viana et al. 2013), suggesting that the statistics of straight lines in this case could be described as essentially resulting from the random addition of straight lines of random lengths. As in the static case discussed above, we also observe here that the evolution of street networks is both qualitatively and quantitatively very different from this biological example and suggest that analogies between cities and living organisms should be treated with great care.

4.2.3 Betweenness centrality impact

Road networks grow by the addition of new streets (edges) and new junctions (nodes) and we discuss here the properties of these new links by looking at their length and centrality. Barthelemy et al. (2013) showed that the average length of new links steadily decreases over time, but more interesting are the features revealed by the BC. In order to evaluate the impact of a new link on the overall distribution of BC in the graph G_t at time t, we first compute the average betweenness centrality of all the links of G_t as

$$\overline{g}(G_t) = \frac{1}{(N(t)-1)(N(t)-2)} \sum_{e \in G_t} g(e) \qquad (4.16)$$

where $g(e)$ is the betweenness centrality of the edge e in the graph G_t. Then, for each new link e^* added in the time window $]t-1,t]$ we consider the new graph obtained by removing the link e^* from G_t (indicated by $G_t \setminus \{e^*\}$). We compute the average edge betweenness centrality for the graph $G_t \setminus \{e^*\}$, and the BC impact

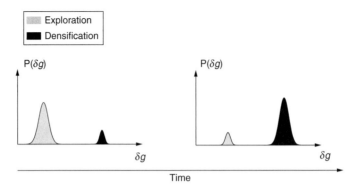

Figure 4.17. Schematic representation of the evolution of the BC impact distribution. In earlier times, exploration was the dominant process while in later times, densification is the most important one.

$\delta g(e^*)$ of edge e^* is then defined as

$$\delta g(e^*) = \frac{\left[\overline{g}(G_t) - \overline{g}(G_t \setminus \{e^*\})\right]}{\overline{g}(G_t)}. \qquad (4.17)$$

In other words, this quantity measures the relative variation of the graph average betweenness due to the removal of the link e^*.

Remarkably, the distribution of this quantity $\delta g(e^*)$ displays two well-separated peaks (for details and figures see Strano et al. 2012), with the first peak tending to increase in time while the second peak decreases, until only one peak remains in the last time-section (1994–2007). These two peaks correspond to very different types of link. The first category consists of links that tend to extend the existing network with dead-ends for example (small δg, first peak), while the second category concerns links that usually connect existing edges to new nodes (large δg, second peak). The distribution of BC impact thus suggests that the evolution of the road network is essentially characterized by two distinct, concurrent processes: one of "densification" (second peak, higher impact on centrality) which corresponds to an increase of local density of the urban texture, and one of "exploration" (first peak, lower impact on centrality) which corresponds to the expansion of the network towards previously non-urbanized areas (see Fig. 4.17). Obviously, since the amount of available land decreases over time, at earlier time–sections (such as in 1833) the fraction of exploration is higher, while in the 1980s it becomes smaller until it completely disappears by 2007. It is important to stress here that even if from the urbanism perspective the main conclusion, with the existence of two main simple processes, is somewhat expected, it is here a non-ambiguous proof that can be reproduced and tested on other datasets. This quantitative, falsifiable aspect is crucial here and distinguishes this approach from the more usual approaches to urbanism.

4.2.4 Evolving patterns of betweenness centrality

As discussed above, the spatial distribution of centrality seems to be a key characterization for spatial networks and the results discussed here seem to confirm this idea.

For the Groane case, we considered the relation between the age of a street and its centrality. Results show that highly central links usually are also the oldest ones. In particular, the links constructed before $t = 1833$ have a much higher centrality than those added in later time-periods. More precisely, we can study the probability $P(b \leq x)$ that a link, appearing at a certain time-period, has a value of betweenness centrality b smaller than or equal to x in the final network at $t = 2007$. We then observe on this quantity (Strano et al. 2012) that the historical structure of oldest links mostly coincides with the highly central links at $t = 2007$. In particular, more than 90% of the 100 most central links in 2007 (and almost 60% of the top 1000) were already present in 1833.

In the case of Paris, we can suspect that Haussmann's impact on congestion and traffic was very important and should therefore be seen on the spatial distribution of centrality. In Fig. 4.18, we show maps of Paris at different times and we indicate the most central nodes such that their centrality $g(i)$ is larger than max g/α with $\alpha = 10$). We can clearly see on this figure that the spatial distribution of the BC is not stable, displays large variations, and is not uniformly distributed over the Paris area (we represented here the node centrality, and similar results are obtained for the edge centrality). In particular, we observe that between 1836 and 1888, Haussmann's work had a dramatic impact on the spatial structure of the centrality, especially near the heart of Paris. Central roads usually persist over time (Strano et al. 2012), but in this case the Haussmann's reorganization acted precisely at this level by redistributing the shortest paths which certainly had an impact on congestion inside the city. After Haussmann the network has largely been stable up to the present day.

It is interesting to note that these maps also provide details about the evolution of the road network of Paris during other periods, which seems to reflect what happened in reality and that can be related to specific local interventions. For example, in the period 1789–1826, between the French Revolution and the Napoleonic empire, the maps shown in Fig. 4.18 display large variations with a redistribution of central nodes, which probably reflects the fact that many religious and aristocratic domains and properties were sold and divided in order to create new houses and new roads. During the period 1826–1836, which corresponds roughly to the beginning of the the July Monarchy, the maps in Fig. 4.18 suggest an important reorganization on the east side of Paris. This seems to correspond to the creation during that period of a new channel in this area (the Saint

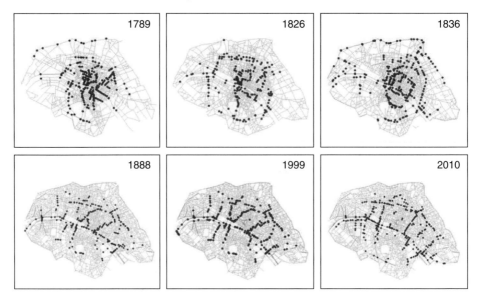

Figure 4.18. Spatial distribution of the most central nodes. We observe for the different periods important reorganizations of the spatial distribution of centrality, corresponding to different specific interventions. In particular, we observe a very important redistribution of centrality during the Haussmann period (1853–1870), with the appearance of a reticulated structure on the 1888 map. Figure from Barthelemy et al. (2013).

Martin channel), which triggered many transformations in the eastern part of the network.

4.2.5 Modeling the road network

Owing to the importance of path-dependence in the evolution of the street pattern, we cannot expect that a model will reproduce the exact layout of a given city. However, we expect a good model to reproduce the large-scale statistical properties of the street pattern, that depends more on the basic mechanisms governing its formation and evolution and less on the large number of individual decisions. In this section we discuss some of the models that were developed along these lines and that might be used as a basis for more evolved ones.

Local optimization

For models relying on local optimization, we assume that new "centers" corresponding to new homes or buildings need to be connected to the existing road network. The rationale for invoking a local optimality principle is then based on costs: every new road is built to connect a center to the existing road network in the most efficient way. The self-organized pattern of streets emerges then as

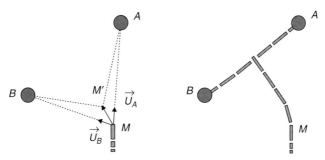

Figure 4.19. M is the network's closest point to both centers A and B. The road will grow to point M' in order to maximally reduce the cumulative distance Δ of A and B from the network. Figure from Barthelemy and Flammini (2008).

a consequence of the interplay of the geometrical disorder and the local rules of optimality. A local optimization model was proposed by Barthelemy and Flammini (2008) and can be illustrated with the simple example shown in Fig. 4.19. We assume that at a given stage of the evolution, two nodes A and B still need to be connected to the network. At any time step, each new node can trigger the construction of a single new portion of road of fixed (small) length. In order to maximally reduce their distance to the network, both A and B would select the closest points M_1 and M_2 in the network as initial points of the new portions of roads to be built. If M_1 and M_2 are distinct, segments of roads are added along the straight lines $M_1 A$ and $M_2 B$. If $M_1 = M_2 = M$, it is not economically reasonable to build two different segments of roads and in this case only one single portion MM' of road is allowed. The main assumption here is that the best choice is to build so as to maximize the reduction of the cumulative distance Δ from M to A and B

$$\Delta = [d(M, A) + d(M, B)] - [d(M', A) + d(M', B)]. \qquad (4.18)$$

The maximization of Δ is done under the constraint $|MM'| = \text{const.} \ll 1$ and a simple calculation leads to

$$\overrightarrow{MM'} \propto \vec{u}_A + \vec{u}_B \qquad (4.19)$$

where $\vec{u}_A$ and $\vec{u}_B$ are the unit vectors from M in the direction of A and B respectively. Rule (4.19) can easily be extended to the situation where more than two centers are required to connect to the same point M. Already in this simple setting non-trivial geometrical features appear. In the example of Fig. 4.19 the road from M will develop a bent shape until it reaches the line AB and intersects it perpendicularly, as commonly observed in most urban settings. At the intersection point, a singularity occurs, with $\vec{u}_A + \vec{u}_B \approx 0$, and one is then forced to build two

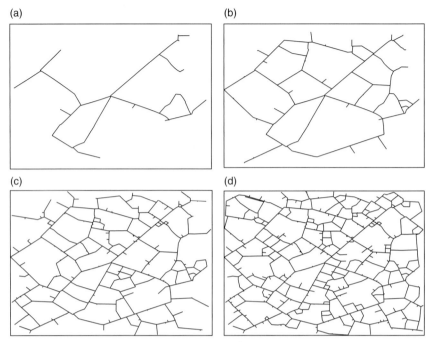

Figure 4.20. Example of the formation of the road network simulated by local optimization growth. Snapshots of the network at different times of its evolution: for (a) $t = 100$, (b) $t = 500$, (c) $t = 2000$, (d) $t = 4000$. In earlier times, we have almost a tree structure and loops appear for larger-density values obtained at later times (the number of loops increases linearly with time). Figure from Barthelemy and Flammini (2008).

independent roads from the intersection to A and B. Interestingly, we note that although the minimum expenditure principle was not used, Rule Eq. (4.19) was also proposed by Runions et al. (2005) in a study about leaf venation patterns.

The growth scheme described so far leads to tree-like structures which are on the one hand economical, but on the other are hardly efficient. For example, the path length along a minimum spanning tree scales as a power 5/4 of the Euclidean distance between the end-points (Duplantier 1989; Coniglio 1989) and better accessibility is gained if loops are present. We can then assume (Barthelemy and Flammini 2008, 2009) that a new node can trigger the construction of more than one portion of road per time step, leading to the existence of loops (see Runions et al. 2005 in the case of leaf venation patterns). This model produces realistic results (an example is shown in Fig. 4.20) that are in good agreement with empirical data (such as degree and shape factor distributions; see Barthelemy and Flammini 2008), which demonstrates that even in the absence of a well-defined blueprint, non-trivial global properties emerge. This agreement confirms that simple local optimization is a good candidate for the main process driving the evolution of city street patterns, but it also shows that the spatial distribution of centers is crucial.

An economical model of network growth

Concerning this aspect about the spatial distribution of centers, a more general model describing the co-evolution of the node distribution $\rho(r)$ and the network structure was proposed by Barthelemy and Flammini (2009). In this model, we assumed that the probability that a new center chooses location i is of the form

$$P(i) \sim e^{\beta z(i)} \quad (4.20)$$

(up to a normalization factor) where β represents the effect of noise (when $\beta = 0$ the choice of location is uniform; and when β is very large the center will choose the maximum of z). The quantity z is in economic terms a composite commodity (see Sections 3.3 and 7.1) and is given by

$$z(i) = Y - C_R(i) - C_T(i). \quad (4.21)$$

The quantity Y is the income (assumed to be the same for all individuals), $C_R(i)$ the renting cost at location i, and $C_T(i)$ is the transportation cost for an individual choosing residence at location i. Simple assumptions were then chosen for these different terms. The transportation cost can be considered as decreasing with the accessibility measured by the BC $g(i)$

$$C_T(i) = B(g_m - g(i)) \quad (4.22)$$

where B and g_m are positive constants (other choices, as long as the cost decreases with centrality, linearly or not, give similar qualitative results). The rent cost is assumed to be proportional to the local population density $\rho(i)$

$$C_R(i) \simeq A\rho(i) \quad (4.23)$$

which is a reasonable assumption as an increasing density will in general lead to a lower offer and higher prices. Places with good accessibility (or BC) will attract a large number of individuals, leading to increasing density and hence increase in renting cost.

Depending on the value of the ratio $\lambda = B/A$, which measures the relative importance of renting cost versus accessibility, we observe the different structures shown on Fig. 4.21. When λ is small, only the density is important and individuals will prefer low-density areas, leading to a uniform state. In contrast, if centrality is the most important factor, individuals will concentrate in areas with good accessibility leading to the formation of local centers. This model is obviously

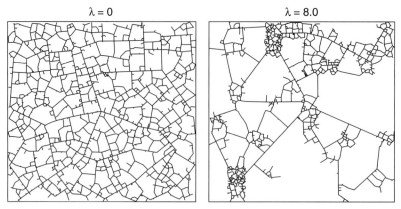

Figure 4.21. Networks obtained for different values of λ. On the left, $\lambda = 0$ (only the density plays a role) and we obtain a uniform distribution of centers. On the right, we show the network obtained for $\lambda = 8$. In this case, the centrality is the most important factor, leading to a few dominant areas with high density. Figure from Barthelemy and Flammini (2009).

very simplified, but offers the possibility of connecting economic considerations (such as optimizing a utility) to the formation of the network and also echoes the discussion about the appearance of a polycentric structure (see Chapter 3).

Finally, another model based on these ideas of optimization was proposed by Courtat et al. (2011). In this model, each new center is added every time step at a certain location and connects to the existing infrastructure network in a way similar to that described by Barthelemy and Flammini (2008). The city generates a spatial field describing the "attractiveness" of every point. This potential field has a hard repulsion term over short distances and a large-distance behavior proportional to the total "mass" of the city (measured by the total length of its roads) and decreasing as a power law with distance. The sum of all influences of all roads then produces local minima and each new node has its own policy. Among others, the parameter ω is related to the "wealth" of the node and controls the number of roads constructed to connect a new settlement: if $\omega = 1$ all possible roads are constructed, and in the opposite case where $\omega = 0$, only the shortest road is built. Another important parameter in this model is the probability of being at the optimal location controlled by the potential. As for Eq. (4.20), for small β the city is "unorganized" and nodes are added at random uniformly, while for large β the probability of sticking to the optimal location is high. When varying the two parameters β and ω, we observe a variety of networks, going from a "favela-like" organization (for large β and small ω) to a highly organized area (for small β), which suggests that this approach could provide an interesting first step for the modeling of urban street networks.

We end this section by noting that there are other models for roads and urban structures and that we focused here on a physicist-like approach with minimal models. In particular, there are many studies on these problems carried out by geographers (usually with the help of cellular automata) and we refer the interested reader to articles in the handbook (Antunes et al. 2009) for more references.

Fragmentation

Street patterns inevitably bring to mind fragmented surfaces, with cracks forming cells of various shapes and sizes. This analogy is not only a nice metaphor, but contains also a part of the real process occurring in urban areas. Larger parcels impose longer detours and for various reasons such as the division of parcels into smaller lots, or the "upgrade" of a trail to a road, it is a process similar to fragmentation (see for example Krapivsky et al. 2010). In the case of urban patterns, the fragmentation process is primarily driven by social forces and the necessity for citizens to reach new buildings or to have efficient routes between important areas of the city. Both problems can be solved by constructing new road segments, which in turn results in the fragmentation of existing blocks and the creation of smaller blocks. However, a study of the patterns of fragments that are drawn by intersecting streets is rarely found in the literature (Lämmer et al. 2006; Fialkowski and Bitner 2008; Nicosia et al. 2016).

Based on this idea of fragmentation, we discuss in this section a class of models for street networks – planar fragmentation models – aimed at reproducing the variety of street patterns observed in cities all around the world (Nicosia et al, 2016). We will focus on the area distribution and refer to Nicosia et al. (2016) for details on other quantities. Generally speaking, the fragmentation process can be divided into different successive steps that are applied at each iteration of the process. At each fragmentation step, we need to specify:

- The cell selection strategy S defined by the probability of selecting a given cell;
- The cutting strategy C. Once the cell and a point are given, we need to specify how we are going to cut the cell.

The specification of these processes defines a class of models which are characterized by different resulting properties of the street network. This class, while very simple in nature, is immensely rich, for many combinations of different strategies can be thought of and implemented. It is therefore necessary to restrict the number of acceptable models.

Intuition tells us that the type of final distribution of fragments depends mainly on the process of block selection, while the exponent of the area distribution might also depend on the cutting strategy. It is therefore important to understand which cell selection processes lead to a power-law distribution of cell area, and which ones don't. We already know (see for example Krapivsky et al. 2010) that randomly selecting the cells with a probability proportional to their surface area does not lead to a pattern with a stable limiting distribution for the area.

In these fragmentation models, we start from a single square cell of area $A = 1$ at time 0 and, at each time step, split a cell into two fragments, so that at time t we will have $t+1$ fragments in total. We denote by $f_t(A)$ the probability density function of the area, representing the number of fragments whose surface area lies in the range $[A, A+dA]$ at time t. The general master equation that we can write for the evolution of $f_t(A)$ reads

$$\frac{\partial f_t(A)}{\partial t} = -S_t(A) f_t(A) + 2 \int_A^1 S_t(B) P_{B \to A} f_t(B) dB \qquad (4.24)$$

where $S_t(A)$ is the probability with which a fragment of size A is chosen at time t to be fragmented, and $P_{B \to A}$ the probability of obtaining a fragment of size A when cutting a fragment of size B. Without loss of generality, we assume that the cutting process is such that $P_{B \to A}$ is uniform, that is $P_{B \to A} = 1/B$. It is easy to prove that if the probability of selecting a cell is proportional to its area, i.e. $S_t(A) \sim A$, then the solution of the master equation is $f_t(A) = t^2 e^{-At}$ and the process does not allow a stationary state (see Krapivsky et al. 2010).

Conversely, if one chooses a uniform distribution $S_t(A) = 1/(t+1)$, the master equation then reads

$$\frac{\partial f_t(A)}{\partial t} = \frac{1}{t+1} \left(-f_t(A) + 2 \int_A^1 \frac{f_t(B)}{B} dB \right). \qquad (4.25)$$

Solving for the limiting stationary probability distribution $f_\infty(A)$, (obtained by imposing $\frac{\partial f_t(A)}{\partial t} = 0$), yields

$$f_\infty(A) = 2 \int_A^1 \frac{f_\infty(B)}{B} dB \qquad (4.26)$$

leading to the differential equation

$$\frac{\partial f_\infty(A)}{\partial A} = -2 \frac{f_\infty(A)}{A} \qquad (4.27)$$

whose solution is

$$f_\infty(A) \sim A^{-2}. \qquad (4.28)$$

Thus the appearance of a power-law distribution of cell areas is compatible with a uniform cell selection strategy.

4.3 Subways

Subway systems have been developed to improve mobility in urban areas – in particular from suburbs to the center – and to reduce congestion. The early history of subways is sometimes connected to large-scale planning, for instance the need to bring population from a growing periphery to the center where production and exchange are usually located. It might seem that subway systems are engineered systems and intentionally structured in a core/periphery shape, with self-organization thus playing only a very minor role. This actually would be true if these subway systems were planned from their beginning to their current shape, but this is not the case for most networks. Their shape results from multiple actions, from planning within a time-limited horizon, set within the wider context of the evolution of the spatial distribution of population and related economic activities. Subway networks actually result from a superimposition of many actions, both at a central level with planning and on a smaller scale with the reorganization and regeneration of economic activity and the growth of residential populations. From this perspective, subway systems are self-organizing systems, driven by the same mechanisms and responding to various geographical constraints and historical paths. This self-organized view leads to the idea that – beside local peculiarities due to the history and topography of the particular system – the topology of world subway networks should display general, universal features, within the limits of the physical geometry and cultural context in which their growth takes place.

4.3.1 All large cities have a subway system

Transportation systems, especially mass transit, are crucial for cities and play an important role in their expansion. In a world where individual transportation increases in cost as cities grow larger, mass transit and, in particular, subway networks are central for the evolution of cities, their spatial organization (Niedzielski and Malecki 2012) and dynamical processes occurring in them (such as epidemic spread, Balcan et al. 2009). The fraction $s(P)$ of cities with population larger than P and with a subway system is shown in Fig. 4.22 (the data were obtained for cities with population larger than 10^5 inhabitants) which confirms that the larger a city, the more likely it is to have some form of mass transit system (see also Daganzo 2010). Approximately 25% of the cities of more than one million individuals have a subway system, 50% of those of more than two million, and all those above 10 million have a subway system (as an indication, an exponential fit of the plot in Fig. 4.22 gives $s(P) = 1 - e^{-P/P_0}$, where the typical population P_0 is of the order 3 million).

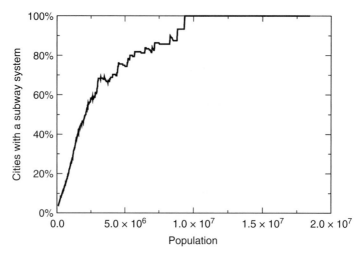

Figure 4.22. Percentage of cities with a subway system versus the population (data from the UN www.unpopulation.org).

4.3.2 Convergence to a universal structure

A important difficulty here is to characterize the space-time evolution of these networks. This is typically a case where the large number of potentially interesting measures might be a problem as it hides the most salient features of the system's evolution. We show here that the introduction of a "template" greatly simplifies the study of these systems (Roth et al. 2012) and provides a guide about the relevant quantities to measure.

We focus on the largest networks in major world cities and ignore currently developing, smaller networks in many medium-sized cities. We thus consider most of the largest metropolitan networks (with at least one hundred stations) that exist in major world cities. These are currently: Barcelona, Beijing, Berlin, Chicago, London, Madrid, Mexico, Moscow, New York City (NYC), Osaka, Paris, Seoul, Shanghai, and Tokyo.

We focus here on urban subway systems and we do not consider longer-distance heavy and light-rail commuting systems in urban areas, such as RER (Réseau Express Régional) in Paris or the overground NetworkRail in London.

In order to get an initial impression of the dynamics of these networks, we first estimate the simplest indicator $v = dN/dt$, which represents the number of new stations built per year. From the instantaneous velocity, we can then compute the average velocity over all years. This average can however be misleading, as there are many years where no stations are built and we describe this by the fraction of "inactivity" time f. The results are shown in Table 4.1 and display some interesting facts. First, it is clear that Shanghai and Seoul are the most recent subway networks, experiencing a rapid expansion that has elevated them to

4.3 Subways

Table 4.1. t_0 is the initial year considered here for the different subway networks. $\bar{v}$ is the average velocity (number of stations built per year), σ_v is the standard deviation of v, and f is the fraction of years of inactivity (no stations built). Table from Roth et al. (2012).

City	t_0	$\bar{v}$	σ_v	f
Beijing	1971	3.3	7.74	79%
Tokyo	1927	2.8	5.47	51%
Seoul	1974	11.2	14.9	20%
Paris	1900	2.6	5.1	60%
Mexico City	1969	3.7	5.9	55%
New York City	1878	3.3	8.3	68%
Chicago	1901	1.9	6.24	71%
London	1863	2.3	3.8	48%
Shanghai	1995	14.9	20.2	31%
Moscow	1936	1.7	1.9	43%
Berlin	1901	1.6	3.3	65%
Madrid	1919	2.3	4.6	59%
Osaka	1934	1.4	4.1	79%
Barcelona	1914	1.4	4.8	78%

amongst the largest networks in the world. For most of these networks the average velocity is in a small range (typically $\bar{v} \in [1.4, 3.7]$) except for Seoul and Shanghai, the more recently developed networks. This is however an average velocity and we observe that:

- for all networks, larger velocities occur at earlier stages of the network;
- large fluctuations occur from one year to another;
- the fraction of inactivity time (the time when no stations are built) is similar for all these networks with an average of about 58%.

We show in Fig. 4.23(left) the time evolution for each city of the number of stations, using an absolute time scale. In particular, the size of the oldest networks seems to progressively reach a plateau. In order to compare the growth across all networks, we compute the average of the number of stations after a certain number of years since their creation. The result is shown in Fig. 4.23(right) and exhibits a linear increase, which indicates that as these networks become large, then for a few decades thereafter new stations represent an increasingly small percentage of existing ones. In other words, the time evolution of all these networks is characterized by small additions and not by sudden, abrupt changes with a large number of stations added in a small time duration. Also, as in the case

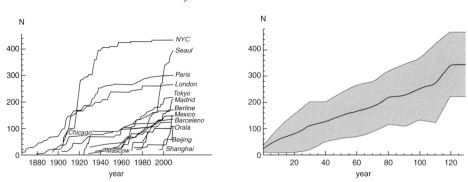

Figure 4.23. (Left) Evolution of the number of stations for various large world subway networks. (right) Evolution of the number of stations y years after creation, averaged over all networks (the gray area marks the standard deviation across all networks). The linear shape indicates that the growth in terms of new stations from one decade to another goes to zero for all these networks, signaling the possible appearance of a stationary limit. Figure from Roth et al. (2012).

of road networks (see previous section), in order to exclude exogenous effects and to be able to compare different networks that experienced different historical developments, we choose the number of stations N as the "clock" and we plot all interesting quantities versus N.

The large subway networks considered here thus converge to a long time limit, where there is always an increasingly smaller percentage of new stations added over time. By inspection, we observe that in most large urban areas, the network consists always of a set of stations delimited by a "ring" that constitutes the "core." From this core, quasi-one-dimensional branches grow and reach out to areas of the city further and further from the core. We note here that the ring, which is defined topologically as the set of core stations which are either at the junction of branches or on the shortest geodesic path connecting these junction stations, may or may not exist as a subway line. For instance, in Tokyo there is such a circular line (the Yamanote line), while for Paris the topological ring does not correspond to a single line. It is interesting to note that the future Grand Paris project (www.societedugrandparis.fr). plans to build large circular rings around the existing Paris area.

More formally, branches are defined as the set of stations which are iteratively built from a terminal station, or a station of degree 1. New neighbors are added to a given branch as long as their degree is 2 – continuing the line, or 3 – defining a fork. In this latter case, the aggregative process continues if and only if at least one of the two possible new paths stemming from the fork is made up of stations of degree 2 or less. Note that the core of a network with no such fork is thus a k-core with $k = 2$ (Seidman 1983). The general structure can then schematically be represented as in

4.3 Subways

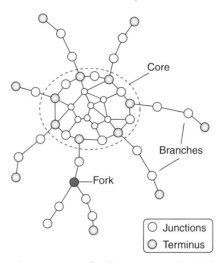

Figure 4.24. Schematic structure of subway networks. A large "ring" encircles a core of stations. Branches radiate from the core and reach further areas of the urban system. The branches are essentially characterized by their size ($\beta(t)$), and their spatial extension (parameter $\eta(t)$). The core is characterized by its average degree ($\langle k_{core}\rangle(t)$), the fraction of nodes of degree 2 (f_2), its number of stations $N_C(t)$ and its size $r_C(t)$. Figure from Roth et al. (2012).

Fig. 4.24. This template suggests what the relevant measures are and in particular how to distinguish the core from the branches. We first characterize this structure with the parameter $\beta(t)$, defined as

$$\beta(t) = \frac{N_B(t)}{N_B(t) + N_C(t)} \qquad (4.29)$$

where $N_B(t)$ and $N_C(t)$ respectively represent the number of stations on branches and the number of stations in the core at time t. We plot in Fig. 4.25(left) this parameter β as a function of N for the networks studied here. It is difficult to draw solid conclusions from this plot, but we can bin these data and represent the average value of β per bin and its dispersion as well (Fig. 4.25(right)). On this figure we may see that the average value of β seems to stabilize slowly to some value in $[0.35, 0.55]$.

The branches have a trivial topological structure (they are essentially 1d lines) and their complexity resides in their spatial structure. We can then determine the average distance (in kms) from the geographic barycenter of the city to all core and branch stations, respectively $\overline{D}_C(t)$ and $\overline{D}_B(t)$ (the barycenter is computed as the center of mass of all stations). In order to characterize the spatial extension of

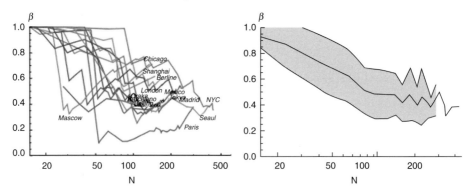

Figure 4.25. (Left) Parameter β as a function of the number of stations N for the different world subways. (Right) Same as (a) but averaged over 20 bins and showing the standard deviation. Figures from Roth et al. (2012).

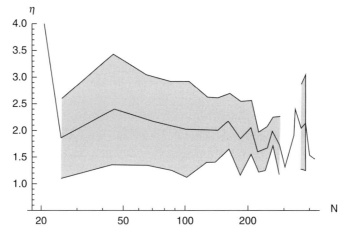

Figure 4.26. Evolution of the ratio η, which characterizes the spatial extension of branches relative to the core. Figure from Roth et al. (2012).

the branches we can compute the ratio $\eta(t)$:

$$\eta(t) = \frac{\overline{D_B(t)}}{\overline{D_C(t)}} \qquad (4.30)$$

which gives a spatial measure of the amount of extension of the branches. We show in Fig. 4.26 the evolution of this parameter with N (the data is binned). This figure shows that in the interval where we have the largest number of subways, the average value of η is around 2, with relatively large fluctuations that seem to decrease with N.

The parameters β and η give an indication of the importance of the core, but do not say anything about its structure. The core is in most cases an almost planar spatial network and can be characterized by many parameters (Barthelemy 2011). It is important to choose those which are not simply related but ideally represent different aspects of the network. At each time step t, we characterize the core structure by its average degree

$$\langle k_{\text{core}}\rangle(t) = \frac{2E_C(t)}{N_C(t)} \tag{4.31}$$

where $N_C(t)$ is the number of core nodes and $E_C(t)$ the number of its edges (we note that this average degree is connected to the standard index $\gamma(t) = E_C(t)/(3N_C(t) - 6)$, where the denominator is the maximum number of links admissible for a planar network; see Haggett and Chorley 1969).

Another simple quantity which describes in more detail the level of interconnections in the core is given by the fraction f_2 of nodes in the core with $k = 2$. For subway systems, nodes with degree larger than 2 are acting as interconnecting stations and the quantity f_2 thus reveals information about the functioning of the system. In the case of a well-interconnected system, this fraction will be small, while sparse cores with few interconnections will have a larger fraction of $k = 2$ nodes.

These two quantities $\langle k_{\text{core}}\rangle$ and f_2 were measured by Roth et al. (2012) and showed that the average degree of the core displays moderate variations around 2.5 (with a slow increase with N). This value is relatively small and indicates that the fraction of connecting stations is also small, meaning that most core stations belong to one single line with few allowing connections. More precisely, for subways with $N < 100$ the fraction of interconnecting stations is increasing with N – which probably corresponds to some organization of the subway – but for larger subways ($N > 100$), the percentage f_2 is increasing again, which probably corresponds to a densification process without the creation of new interconnections.

Branches and fractals

The branches constitute probably the most interesting part of these large networks. They are made of essentially one-dimensional lines that explore space at a large distance from the center. We first consider the number $\mathcal{N}_B$ of these branches and propose a simple argument for estimating this quantity. If we assume that the distance between the starting points from the ring of two consecutive branches is constant (which probably corresponds to an optimum in terms of distribution and cost), the number of branches is then proportional to the perimeter of the core structure. The core is a relatively dense planar graph and contains a number of nodes proportional to N, and its perimeter thus scales as $\sqrt{N}$. These arguments

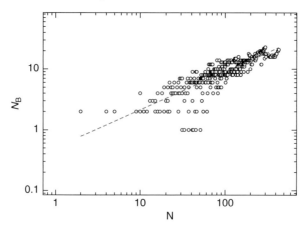

Figure 4.27. Log–log plot of the number of different branches versus the number of stations for the different subway networks considered in Roth et al. (2012). The dashed line is a power-law fit with exponent ≈ 0.6. Figure from Roth et al. (2012).

thus lead to the following scaling for the total number of branches

$$\mathcal{N}_B \sim \sqrt{N}. \tag{4.32}$$

We display the number of branches versus the number of stations N for the various networks considered here. A power-law fit of the data presented in Fig. 4.27 gives $\mathcal{N}_B \sim N^b$, with $b \approx 0.6$ ($r^2 = 0.85$) consistent with this simple argument.

Following earlier studies on the fractal aspects of subway networks (Benguigui 1992), we consider the number of stations $N(r)$ at a distance less than or equal to r, where the origin of distances is the barycenter of all stations. In order to compare the different networks, we first determine the size r_C of the core and N_C its number of nodes for the last year available at the time of this study (2009). We then rescale r by r_C and $N(r)$ by N_C and the results are shown in Fig. 4.28. This figure displays several interesting features. First, the short distance regime $r < r_C$ is well described by a behavior of the form $N(r) \sim \rho_C \pi r^2$, consistent with a uniform density ρ_C of core stations. For very large distances, we observe for most networks a saturation of $N(r)$. The interesting regime is then for intermediate distances

$$r_C \ll r \ll r_{\max} \tag{4.33}$$

where $r_{\max}$ is the maximum branch size. This intermediate regime is characterized by an exponent that depends on the system considered. A similar result was obtained earlier by Benguigui (1995) where the authors observed for Paris that $N(r > r_c) \sim r^{0.5}$, a result that was at that time difficult to understand in the framework of fractal geometry. Instead, these regimes can be easily understood in terms of the core-and-branches model, with the additional ingredient that the

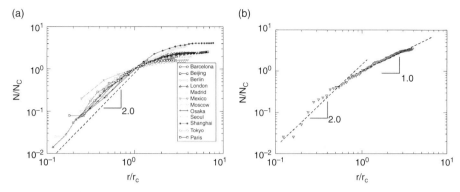

Figure 4.28. (a) Rescaled number of stations at distance r from the barycenter as a function of the rescaled variable r/r_C where r_C is the size of the core defined as $N(r = r_C) = N_C$ (shown here in log–log). The dotted line represents a power law $\sim r^2$ and serves as a guide to the eye. (b) Case of Moscow, where the two regimes ($r < r_C$ and $r > r_C$) with their different exponents are visible (the dotted lines serve here as a guide to the eye). Figure from Roth et al. (2012).

spacing $\Delta(r)$ between consecutive stations increases with r. Within this picture (and assuming isotropy), $N(r)$ is given by

$$N(r) \sim \begin{cases} \rho_C \pi r^2 & \text{for } r < r_C \\ \rho_C \pi r_C^2 + \mathcal{N}_B \int_{r_C}^{r} \frac{dr}{\Delta(r)} & \text{for } r_C < r < r_{\max} \\ N & \text{for } r > r_{\max} \end{cases} \quad (4.34)$$

where N is the total number of stations, $\mathcal{N}_B$ is the number of branches and $\Delta(r)$ is the average spacing between stations on branches at distance r from the barycenter.

In order to test these simple expressions, we determine the various parameters – namely $\mathcal{N}_B$, N_C, r_C, and $\Delta(r)$ – and plot the resulting shape of Eq. (4.34) against the empirical data. It is easy to determine empirically the numbers $\mathcal{N}_B$, N_C, and r_C, but the quantity $\Delta(r)$ is extremely noisy due to the small number of points (all these numbers are determined for the year 2009), especially for large values of r close to $r_{\max}$, where there is often no more than a handful of stations.

In general, the intermediate distance behavior $r_C < r < r_{\max}$ is of the form

$$N(r_C < r < r_{\max}) \sim r^{1-\tau} \quad (4.35)$$

where τ denotes the exponent governing the interspacing decay $\Delta(r) \sim r^\tau$. For most networks, the regime $r_C < r < r_{\max}$ is small and as already mentioned $\Delta(r)$ is very noisy. Rough fits in different cases give a behavior for Eq. (4.34) consistent with data.

A less noisy situation is obtained in the case of Moscow which has long branches and a roughly constant interstation spacing. In this case we observe for $r > r_C$ a behavior of the form $N(r) \sim \mathcal{N}_B r$ (see Fig. 4.28b). For the other networks, we observe an increasing trend but an accurate estimate of τ is difficult to obtain. For example, a fit over a decade of data gives for Paris $\tau \approx 0.5$ (with $r^2 = 0.74$), in agreement with the result obtained by Benguigui (1995). Despite the difficulty of obtaining accurate quantitative results, the simple picture of a core and radial branches seems to provide a simple explanation of empirical observations.

4.3.3 Scaling and modeling for subways

Many studies (Kansky 1963; Derrible and Kennedy 2009; Levinson 2012) explored the interplay between regional characteristics and the structure of transportation networks, and here we discuss a simple picture relating the network's most basic quantities and the region's properties. This large-scale framework, based on a cost-benefit analysis (Black 1971; Louf and Barthelemy 2013), allows us to understand how subway networks (and in the next section railroads) scale with some of the substrates' most basic attributes.

A transportation network is at least characterized by its total number of stations, its total length, and the total yearly ridership. On the other hand, a city is characterized by its area, its population and its Gross Domestic Product (GDP). Because transportation systems do not grow in empty space, but result from multiple interactions with the substrate, it is important to relate the network characteristics and socio-economical indicators.

We consider an iterative growth where at each step an edge e is built such that the cost function

$$Z(e) = B(e) - C(e) \qquad (4.36)$$

is maximum. The quantity $B(e)$ is the expected benefit and $C(e)$ the cost of edge e. We consider networks after they have been built, and we assume that they are in a "steady state" for which we can write a cost function of the form

$$Z = \sum_e Z(e) = B - C \qquad (4.37)$$

where B is the total expected benefits and C the total cost, mainly due to maintenance (in the steady-state regime). We further assume that, during this steady state, operating costs are balanced by benefits, which reads

$$Z \approx 0. \qquad (4.38)$$

Indeed, because lines and stations cost money to be maintained, we expect the network to adapt to the way it is being used. We can therefore reasonably expect

4.3 Subways

that "at lowest order" the cost of operating the system is compensated by the benefits gained from its use.

The estimate for Z depends obviously on the nature of the network and, as we will see in the next section, is very different for subways and railroads. For subways, the total benefits in the steady-state regime are simply connected to the total ridership per unit time R and the ticket price f. The costs, on the other hand, are due to the maintenance costs of the lines and stations, so that we can write (for a given period of time)

$$Z_{\text{subway}} = R f - \epsilon_L L - \epsilon_S N_s \tag{4.39}$$

where L is the total length of the network, ϵ_L the maintenance cost per unit length and unit time of a line, N_S the total number of stations and ϵ_S the maintenance cost per unit time of a station.

The ridership of a system, given its characteristics and those of the underlying city, is difficult to estimate and we discuss here a simple argument (Louf et al. 2014). Very generally, the number R_i of people using the station i is a function of the area C_i serviced by this station – the "coverage" (Derrible and Kennedy 2009) – and of the average population density $\rho = \frac{P}{A}$ in the city

$$R_i = \xi_i C_i \rho. \tag{4.40}$$

The quantity ξ_i is a random number of order one representing the fraction of people who are in the area serviced by the station and who use the subway. The coverage depends *a priori* on local particularities such as the accessibility of the station, and varies from one station to another. We take here a simple approach and assume that on average it is given by

$$C_i \sim \pi d_0^2, \tag{4.41}$$

where d_0 is the typical size of the attraction basin of a given station. If we assume that d_0 is constant, the total ridership can be written as

$$R = \sum_i R_i \sim \overline{\xi} \pi d_0^2 \rho N_s, \tag{4.42}$$

where $\overline{\xi} = \frac{1}{N_s} \sum_i \xi_i$ is of the order of 1.

On Fig. 4.29(left) we show the ridership R as a function of $N_s \rho$ for 138 metro systems across the world and we observe that the data are consistent with a linear behavior. The measure of the corresponding slope gives 800 km²/year, leading to the estimate $d_0 \approx 500$ m. We illustrate this result on Fig. 4.29(right) by representing each subway stations of Paris with a circle of radius 500 m.

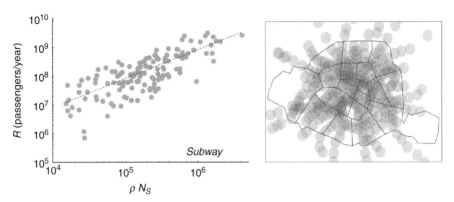

Figure 4.29. Relationship between ridership and coverage. (Left) We plot the total yearly ridership R as a function of ρN_s. A linear fit on the 138 data points gives $R \approx 800 \, \rho N_s$ ($r^2 = 0.76$), which leads to a typical effective length of attraction $d_0 \approx 500$ m per station. (Right) Map of Paris (France) with each subway station represented by a circle of radius 500 m. Figure taken from Louf et al. (2014).

So far, the distance d_0 appears here as an intrinsic feature of users' behaviors: it is the maximal distance that an individual would walk to go to a subway station. We can relate this distance to the average interstation distance ℓ_1, which is another chacteristic scale of the subway system. By definition, this distance depends on the average degree $<k>$ of the network so that $\ell_1 = \frac{2L}{N_s <k>}$. It has been found (Roth et al. 2012) that for the 13 largest subway systems in the world, $<k> \in [2.1, 2.4]$, so that we can reasonably take $<k>/2 \approx 1$ and thus

$$\ell_1 \simeq \frac{L}{N_s}. \tag{4.43}$$

We expect that for a properly designed system ℓ_1 will match human constraints. Indeed, if $d_0 \ll \ell_1$, the network is not dense enough and in the opposite case $d_0 \gg \ell_1$, the system is not economically interesting. We can thus reasonably expect that the interstation distance is of the order twice d_0

$$d_0 = \frac{\ell_1}{2} = \frac{L}{2 N_s}. \tag{4.44}$$

It follows from this assumption that the interstation distance is constant and independent of the population size, and in order to test this, we plot on Fig. 4.30 (left) the total length of subway networks as a function of the number of stations. The data agree well with a linear fit $L \sim 1.13 \, N_S$ ($r^2 = 0.93$). We also plot on Fig. 4.30 (right) the normalized histogram of the inter station length, showing that the interstation distance distribution is peaked around an average value $\overline{\ell_1} \approx$

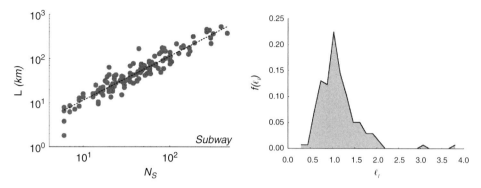

Figure 4.30. Relation between the length and the number of stations. (Left) Length of 138 subway networks in the world as a function of the number of stations. A linear fit gives $L \sim 1.13 N_S$ ($r^2 = 0.93$) (Right) Empirical distribution for the 138 subway networks in the world of the interstation length. The average inter station distance is found to be $\overline{\ell_1} \approx 1.2$ km and the relative standard deviation is approximately 440 m. Figure from Louf et al. (2014).

1.2 km with a variance $\sigma \approx 400$ m, consistently with the value found above for $d_0 \approx 500$ m.

We can then express ℓ_1 in terms of the system's characteristics. Indeed, the total ridership now reads

$$R \sim \bar{\xi} \pi \rho \frac{L^2}{N_S}, \qquad (4.45)$$

and in the steady state $Z_{\text{subway}} \approx 0$ (Eq. 4.38) leads to the relation between the total length of the network and the number of stations (at first order in ϵ_s/ϵ_L)

$$L \sim \left(\frac{4\epsilon_L}{\pi \xi f \rho} + \frac{\epsilon_s}{\epsilon_L} \right) N_s, \qquad (4.46)$$

The interstation distance then reads

$$\ell_1 = \frac{4\epsilon_L}{\pi \xi f \rho} + \frac{\epsilon_s}{\epsilon_L}. \qquad (4.47)$$

This relation implies that the interstation distance increases with the station maintenance cost, and decreases with increasing line maintenance costs, density and fare. The adjustment of ℓ_1 to match $2d_0$ can then be made through the fare price or subsidies by the local authorities or national government.

So far, we have a relation between the total length L and the number of stations N_s, but we need another equation in order to compute their value. Intuitively, it is clear that the number of stations – or equivalently the total length – of a subway system is an increasing function of the wealth of the city. We thus assume a simple,

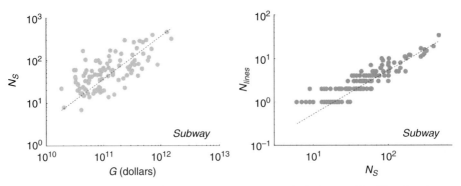

Figure 4.31. (Left) Size of the subway system and city's wealth. We plot the number of stations for the different subway systems in the dataset as a function of the Gross Metropolitan Product of the corresponding cities (obtained for 106 subway systems). A linear fit (dashed line) gives $N_s = 2.51\, 10^{-10}\, G$ ($r^2 = 0.73$). (Right). We plot the number of metro lines N_{lines} as a function of the number of stations N_s. A linear fit on the 138 data points gives $N_{lines} \approx 0.053\, N_s$ ($r^2 = 0.93$), or, in other words, metro lines comprise on average 19 stations. Figures from Louf et al. (2014).

linear relation of the form

$$N_s = \beta \frac{G}{\epsilon_s} \qquad (4.48)$$

where G is the city's Gross Metropolitan Product (GMP), and β the fraction of the city's wealth invested in public transportation. This relation can equivalently be interpreted as the proportional relation between the number of stations per person and the city's development, as measured by its GMP per capita. In Fig. 4.31 (left) we plot the number of stations of different metro systems around the world as a function of the Gross Metropolitan Product of the corresponding city. A linear fit agrees relatively well with the data ($r^2 = 0.73$, dashed line), and gives $\frac{\epsilon_s}{\beta} \approx 10^{10}$ dollars/station. The dispersion around the linear average behavior is however important: more data are needed in order to investigate whether differences in the construction costs and investments (or the age of the system) can explain the dispersion, or if other important parameters need to be taken into account. We also consider the number of different lines with distinct tracks. A natural question is how the number of lines N_{lines} scales with the number stations N_s. We show the number of lines as a number of stations in Fig. 4.31 (right) and observe that the data agree with a linear relationship between both quantities ($r^2 = 0.93$). In other words, the number of stations per line is distributed around a typical value of 19, whatever the size of the system.

Finally, we note that that this framework does not explain the appearance of subway networks. We observe empirically that the GDP of cities that have a subway system is always larger than about 10^{10} dollars, an interesting fact that calls for a theoretical explanation and deserves further studies.

4.4 Digression: Railroads

4.4.1 Scaling

In the subway case, we saw that the interstation distance ℓ_1 is such that it matches human constraints: $\ell_1 \sim 2d_0$, where d_0 is the typical distance that one would walk to reach a subway station. For the railroad network, the logic is however different: while subways are built to allow people to move within a dense urban environment, the purpose of a railroad is to connect different cities in a country. Owing to long distances and hence high costs, it seems reasonable to assume that each city is connected to its closest neighboring city. The railroad network then appears as a planar graph connecting, in an economical way, randomly distributed cities in the plane. For a country of total area A and with N_s train stations, the typical distance ℓ_N between nearest stations is then

$$\ell_N = \sqrt{\frac{A}{N_s}}, \tag{4.49}$$

and the total length $L \sim N_s \ell_N$ is given by

$$L \sim \sqrt{A N_s}. \tag{4.50}$$

In order to test this relation for different countries, we show the adimensional quantity $\frac{L}{\sqrt{A}}$ as a function of the number of stations N_s on Fig. 4.32. A power-law fit gives an exponent 0.50 ± 0.08 ($r^2 = 0.87$), which is consistent with the simple argument given above.

Owing to distances involved, the ticket price for railroads usually depends on the distance traveled and we denote by f_L the ticket price per unit distance. The relevant quantity for benefits is therefore not the raw number of passengers – as in subways – but rather the total distance traveled on the network T. Also, again due to the long distances spanned by the network, the costs of stations can be neglected as a first approximation, and we get

$$Z_{\text{train}} \simeq T f_L - \epsilon_L L. \tag{4.51}$$

In the steady-state regime $Z_{\text{train}} \approx 0$, or in other words the revenue generated by the network use must be of the order of the total maintenance costs, which leads to

$$T \sim \frac{\epsilon_L}{f_L} L. \tag{4.52}$$

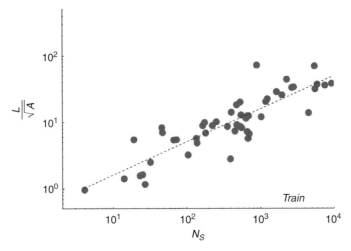

Figure 4.32. Total length of the national railroad network L rescaled by the typical size of the country $\sqrt{A}$ as a function of the number of stations N_s. The dashed line shows the best power-law fit on the 50 data points with an exponent 0.50 ± 0.08 ($r^2 = 0.87$). Figure from Louf et al. (2014).

In addition, if we assume that the order of magnitude of a trip is given by ℓ_N, the total traveled length is simply proportional to the ridership $T \sim \ell_N R$ leading to

$$R \sim \frac{\epsilon_L N_s}{f_L}. \quad (4.53)$$

We show in Fig. 4.33(left) the total daily ridership R as a function of the total number of stations N_s and, despite the small number of available data points, a linear relationship between both these quantities seems to agree with empirical data on average ($r^2 = 0.86$). This result should however be taken with caution owing to the important dispersion that is observed around the average behavior, and the small number of observations.

We now would like to understand what property of the underlying country determines the total length of the network, and why networks are longer in some countries than in others. *A priori*, when estimating the cost of a railroad network, one should take into account both the costs of building lines and stations. Considering the distances involved, the cost of building a station is negligible compared to that of building the actual lines and we thus can reasonably expect to have

$$L \sim \frac{\alpha G}{\epsilon_L}, \quad (4.54)$$

where G is here the country's Gross Domestic Product (GDP), used as an indicator of the country's wealth, and $\alpha < 1$ is the ratio of the GDP invested in railroad

4.4 Digression: Railroads

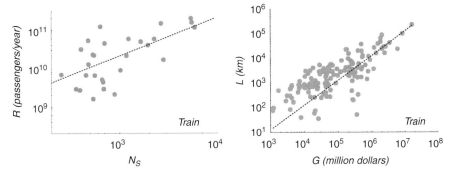

Figure 4.33. (Left) The total yearly ridership R of the railroad networks as a function of the number of stations. A linear fit on the 47 data points gives $R \sim 7.0\,10^8\,N_s$ ($r^2 = 0.86$). (Right) Total length of the railroad network L as a function of the country GDP G. The dashed line shows the linear fit on the 138 data points, which gives $\epsilon_L/\alpha \approx 10^4$ dollars.km^{-1} ($r^2 = 0.91$). Figures from Louf et al. (2014).

Table 4.2. Summary of the differences between subways and railroads. N_s is the number of stations, L the total length of the network, P the population of the region considered (cities for subways and countries for railroads) and A its area. The quantity R is the total ridership and the wealth of the region is described by the GDP G. Table from Louf et al. (2014).

Quantity	Subway	Train
L/N_s	cste.	$\sqrt{\frac{A}{N_s}}$
R	$\frac{P}{A}N_s$	N_s
G	N_s	L

transportation. We show L as a function of G on Fig. 4.33(right) and the data agree well ($r^2 = 0.91$) with a linear dependence between L and G. Again, the dispersion indicates that the linear trend should only be understood as an average behavior and that local particularities can have a strong impact on the important deviations observed.

4.4.2 Are subways and railroads the same?

In these previous two sections, we observed scaling relations for global properties of railroads and subways, and the existence of such relations suggests that basic, common mechanisms are at play during their evolution. A probable reason for the presence of these systems is the mobility demand and their structure is driven by economic mechanisms that seem to be the same for all countries, independently from any cultural or historical considerations. The fact that macroscopic properties seem to be independent from specific details opens the possibility for simple modeling, and we saw here how a general framework allows to connect the properties of railroad and subway systems (ridership, total length and number of stations) to the socio-economic and spatial characteristics (population, area, GDP) of the country or city where they are built. Despite their simplicity, these arguments agree satisfactorily with data gathered for almost 140 subway systems and 50 railroad networks accross the world. The main result is that the knowledge of simple characteristics of a country or a city is enough to give an estimate of the size and use of its transportation system.

These results suggest also that the fundamental difference between railroads and subways lies in the determination of the interstation distance. While it is imposed by human constraints in the subway case, the railroad network in contrast has to adapt to the spatial distribution of cities in a country. This remark is at the heart of the different behaviors observed for railroads and subways that are summarized in Table 4.2.

These simple arguments are able to explain the average behavior of various quantities, but it would be nevertheless interesting to identify deviations from these behaviors, and check if they can be explained by topological properties of the system, as suggested by Derrible and Kennedy (2009) or by Levinson (2012) in the case of road networks. The advantage of the simple cost-benefit framework discussed here is that it can serve as a good starting point for enabling the integration of other ingredients and mechanisms. This framework could also serve as a null model in order to quantify the efficiency of specific transportation systems.

5
Mobility patterns

Mobility is obviously a crucial phenomenon in cities. In fact, it is probably one of the most important mechanisms that govern the structure and dynamics of cities. Indeed, individuals go to cities to buy, sell or exchange goods, to work, or to meet with other individuals and for this they need different transportation means. This is where technology enters the problem through the (average) velocity of transportation modes. This average velocity increased during the evolution of technology and modified the structure and organization of cities. For example, we see in Fig. 5.1 that the "horizon" of an individual depends strongly on her transportation mode. For a walker, the horizon is essentially isotropic and small, while the car allows for a wider exploration but one which is anisotropic and follows transportation infrastructures. This correlation between the spatial structure of the city and the available technology at the moment of its creation is clearly illustrated by Anas et al. (1998) for US cities. Many major cities, such as Denver or Oklahoma City, developed around rail terminals that triggered the formation of central business districts. In contrast, automobile-era cities that developed later, such as Dallas or Houston, have a spatial organization that is essentially determined by the highway system.

In terms of mobility, the city center is also the location that mimimizes the average distance to all other locations in the city. Very naturally, it is then the main attraction for businesses and residences, which leads to competition for space between individuals or firms, giving rise to the real-estate market. There is also a well-known relation between land-use and accessibility, as was discussed some time ago by Hansen (1959), and new, extensive datasets will certainly enable us in the future to characterize precisely the relation between these important factors.

It is of course very difficult to make an exhaustive review about all studies on mobility and we will focus in this chapter on several specific points. We will mostly describe the general features of mobility and will leave the discussion about multimodal aspects for Chapter 6. First, we consider the central quantity in these

130   *Mobility patterns*

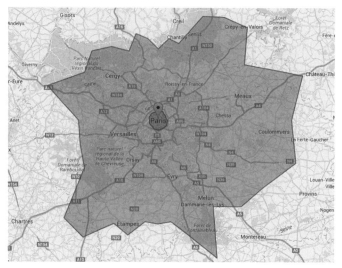

Figure 5.1. Influence of the transport technology: comparison of the area that can be reached in one hour from the center of Paris (France) by walking (darker area) or by car (light gray). Map from Google and figure from www.oalley.fr

studies – the origin-destination (OD) matrix – and we discuss how to extract useful information from it. In particular, we present a method that allows comparison between different OD matrices for different cities. After having discussed some empirical facts, we turn to the theoretical description of mobility in urban systems. We start by describing the gravity model and we then focus on more recent approaches such as the radiation model or the closest opportunity model.

5.1 Typology of origin–destination matrices

In transportation research and urban planning, the daily mobility of individuals is usually captured by Origin–Destination (OD) matrices, which contain the flows of individuals going from one area to another (de Dios Ortúzar and Willumsen 1994; Weiner 1999). An OD matrix thus encapsulates the complete information about flows of individuals in a city, at a given spatial scale and for a specific purpose.

The OD matrix was until recently obtained from surveys that are necessarily incomplete and inaccurate. With new data sources such as cell phones or GPS, for example, we can now have access to real-time, high-resolution OD matrices. For cell phones we know the location of the nearest communication antenna (the Base Transceiver Station, or BTS) to which phones are connecting. Each call detail record (CDR) then contains various bits of information such as the time of the call, its duration and the IDs of both antennae (for the caller and the callee). Applying statistical filters to this dataset allows us to identify the residence and activity

5.1 Typology of origin–destination matrices 131

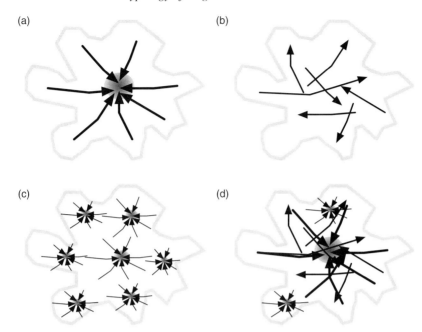

Figure 5.2. Generic types of mobility patterns according to Bertaud and Malpezzi (2003). (a) Monocentric pattern, (b) random flows, (c) urban village structure with many subcenters having their own basin of attraction, and (d) a mixture of the previous types. Figure inspired by Bertaud and Malpezzi (2003).

locations, and to discuss mobility and urban features (Louail et al. 2014, 2015; Calabrese et al. 2015).

This information about the OD matrix enables us to quantitatively characterize mobility. In particular, Bertaud and Malpezzi (2003) proposed various generic forms of mobility patterns that are depicted in Fig. 5.2. We recognize for example the typical pattern for a monocentric city, random flows, or a polycentric structure, and one can wonder whether these drawings are representative of what really happens in our cities.

In order to test these patterns with data obtained for the OD matrices, we have to face a new problem: these OD matrices are large objects and cannot be readily used for identifying simple, dominant patterns. The extraction of a clear and simple footprint of the structure of large, weighted and directed networks is a difficult problem and we need a method that allows comparison between OD matrices for a given city at different times or for different cities. Such coarse-grained information can lead to stylized facts that encode the essence of the mobility phenomenon and form the basic element to guide theoretical modeling. This method can also be useful for validating synthetic results of urban mobility models (such as in Eubank et al. 2004, for example), and for comparing different models. An accurate modeling of mobility is indeed crucial in a large number of applications, including the important case of epidemic spreading which needs to be better understood, in particular at the intra-urban level.

In this section, we discuss such a method proposed by Louail et al. (2015), which extracts a coarse-grained signature of mobility networks, in the form of a 2×2 matrix that separates the mobility flows into four categories.

5.1.1 Extracting coarse-grained information from OD matrices

The OD matrix is a large matrix and as such does not provide clear, synthetic and useful information about the structure of the mobility in the city. More precisely, an OD matrix $\mathbf{F}$ is an $n \times m$ matrix, where n is the number of different "Origin" zones, m is the number of "Destination" zones, and F_{ij} is the number of people commuting from place i to place j during a given period of time (in many practical applications $n = m$). We thus need to define zones implying the presence of a spatial scale at which the mobility data has been collected and aggregated (for example the size a of a grid, Fig 5.3). Traditionally the zones that are used to partition the city are administrative units, whose size can vary from census and electoral units to whole departments or states, depending on the purpose for building the OD matrix. In the case of cell phones, the natural division is given by the Voronoi tesselation based on antennas (see for example Louail et al. 2014).

A particular case of OD matrices concerns the "commuting networks" in cities, where edges represent flows of individuals who travel daily from their residential neighborhood to their main activity area. Several types of links can be distinguished in these mobility networks, some of them constituting the backbone of the city by connecting major residential neighborhoods to employment centers,

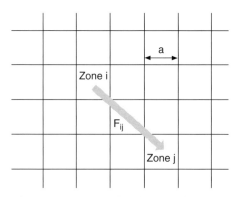

Figure 5.3. Schematic representation of an OD matrix. The area (city or a larger zone) is divided into zones characterized by a size a. The OD matrix element from i to j is the number of individuals going from i to j in a given period of time.

5.1 Typology of origin–destination matrices

while other flows converge from smaller residential areas to important employment centers, or diverge from major residential neighborhoods to smaller activity areas. In addition, the spatial properties of these commuting flows are fundamental in cities and a relevant method should be able to take this aspect into account.

We thus assume that for a given city, we know the OD matrix F_{ij} which represents the number of individuals living in the location i and commuting to the location j where they have their main, regular activity such as work or school. The diagonal of this matrix represents individuals who are "immobile" on this scale and can thus be omitted for a mobility analysis.

In order to extract a simple signature of the OD matrix, we first identify both the residential and work locations that have an exceptionally large density of individuals, the "hotspots" (see Section 3.2). The number of residents of the cell i is given by $\sum_{j \neq i} F_{ij}$, its number of workers is given by the "incoming" flow $\sum_{j \neq i} F_{ji}$, and the hotspots will correspond to local maxima of these quantities. Once we have determined the cells that are the residential and the work hotspots (some cells can possibly be both), we separate the flows according to the type of both their origin and destination (hotspot or non-hotspots) The OD matrix then becomes a 4-quadrants matrix where we sum the number of commuters and normalize it by the total number of commuters in the OD matrix, which gives the proportion of individuals in each of the four categories of flow. We thus reduce the OD matrix to a 2×2 matrix of the form

$$\Lambda = \begin{pmatrix} I & D \\ C & R \end{pmatrix} \tag{5.1}$$

where

$$I = \sum_{i \in H_R, j \in H_W} F_{ij} / \sum_{i,j} F_{ij} \tag{5.2}$$

$$C = \sum_{i \in \overline{H_R}, j \in H_W} F_{ij} / \sum_{i,j} F_{ij} \tag{5.3}$$

$$D = \sum_{i \in H_R, j \in \overline{H_W}} F_{ij} / \sum_{i,j} F_{ij} \tag{5.4}$$

$$R = \sum_{i \in \overline{H_R}, j \in \overline{H_W}} F_{ij} / \sum_{i,j} F_{ij} \tag{5.5}$$

where H_R and H_W are the sets of residential and workplace hotspots and $\overline{H_R}, \overline{H_W}$ are the complementary sets. The different terms correspond to:

- I is the proportion of **I**ntegrated flows that go from residential hotspots to work hotspots;

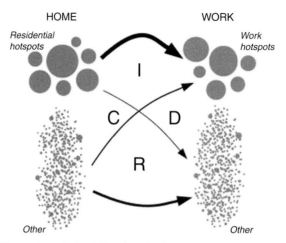

Figure 5.4. Illustration of the ICDR method. The commuting flows are decomposed into four categories: the Integrated flows (I) from hotspot to hotspot, the Convergent flows (C) to hotspots, the Divergent flows (D) originating at hotspots and finally the Random flows (R), which neither start nor end at hotspots. Figure taken from Louail et al. (2015).

- C is the proportion of **C**onvergent flows that go from random residential places to work hotspots;
- D is the proportion of **D**ivergent flows that go from residential hotspots to random activity places;
- R is the proportion of **R**andom flows that occur "at random" in the city, i.e. that are going from and to places that are not hotspots.

By construction, the numbers $I, C, D, R \in [0; 1]$ and are such that $I + C + D + R = 1$. This matrix Λ is thus a very simple footprint of the OD matrix that gives an expressive picture of the structure of commuting in the city, as illustrated by Fig. 5.4.

5.1.2 Comparing mobility networks

The method presented above produces then a reduced version of the OD matrix which allows to analyze the mobility structure of a city and to compare different cities. This is particularly important, as there is an important literature in quantitative geography and transportation research that focuses on the morphological comparison of cities (Tsai 2005; Guérois and Pumain 2008; Schwarz 2010; Le Néchet 2012). Most of these comparisons are based on morphological indicators such as built-up areas, residential density, number of sub-centers, etc. (Tsai 2005; Guérois and Pumain 2008), and aggregated mobility indicators such as motorization rate, average number of trips per day, energy consumption per capita

5.1 Typology of origin–destination matrices

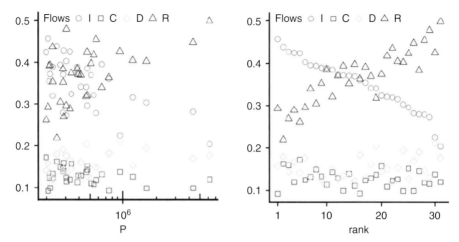

Figure 5.5. Results of the ICDR method for 31 Spanish cities. (Left) I (integrated), C (convergent), D (divergent) and R (random) values versus population size for 31 Spanish urban areas. (Right) Same ICDR values as in the left panel but sorted by decreasing order of I (note that by definition, we have for each city $I + C + D + R = 1$). It is remarkable that I and R dominate and seem almost sufficient to distinguish cities, while C and D are almost constant whatever the city size. Figures taken from Louail et al. (2015).

per transport mode, etc. (Schwarz 2010; Le Néchet 2012). These studies focused on the spatial organization of residences and workplaces but didn't consider the spatial structure of commuting trips and the organization of commuting flows. This is where this "reduction" method can bring new information about the dynamics of commuting in cities.

We illustrate this method using data extracted from cell-phone records in thirty-one Spanish urban areas during a five-week period (Louail et al. (2015)). The hotspots are determined using the parameter-free method described in Section 3.2. The origins are determined by residences and destinations by workplaces, which implies that there are hotspots for both categories, residences and workplaces. Once these hotspots are determined, the ICDR values entering the matrix in Eq. (5.1) are computed and, for the 31 Spanish urban areas considered in this study, the values obtained are shown in Fig. 5.5. In Fig. 5.5(left), these values are plotted versus the population size of these cities. We observe that the proportion I of individuals that commute from hotspot to hotspot decreases as the population size increases, while the proportion R of "random" flows increases; and that the proportions C and D of convergent and divergent flows seem surprisingly constant whatever the city size. In Fig. 5.5(right) the same values are sorted by decreasing values of I, show clearly that the I and R values are the relevant parameters for distinguishing cities from each other. The decay of the I flows in favor of R flows when P increases shows that population growth among Spanish cities is coupled with a delocalization of both activity places and residences. As cities become bigger, the number of residential and employment hotspots grows (sublinearly; see Section 3.2), but these hotspots attract a smaller part of the commuting flows.

These results can be compared to a random null model where flows are uniformly distributed among nodes (keeping the number of outgoing and incoming individuals fixed) and shows that the $ICDR$ signatures of cities are characteristic of their commuting structure (Louail et al. 2015). In particular, it appears that the larger a city, the less random it appears. This is in contrast to the naïve expectation that the larger a city, the more disordered is the structure of individuals' mobility. For large cities, there seems to be a commuting backbone which cannot result from purely random movements of individuals, and that is not prevalent between hotspots. This backbone is the footprint of the city's structure and history, and probably results from strong constraints and efficiency considerations.

It is also interesting to note that the different types of flow also correspond to different spatial scales. The average distance for all categories of flow increases with population size, an expected effect as the city's area also grows with population size. We also observe that the average distance of convergent flows C increases faster than for other types of flow (in particular, faster than the average distance associated with divergent flows D, showing that these flows are not symmetric as one could have naïvely expected). This result means that for this set of Spanish cities, commuters from small residential areas to important activity centers travel on average a longer distance than all other individuals. This observation could be an indication that for this set of cities, residential areas have expanded while activity centers remained at their location, leading to longer commuting distances.

Classification of cities

Finally, the $ICDR$ signature of their OD matrix enables us to cluster cities with respect to the structure of their commuting patterns. By introducing a distance between cities based on their $ICDR$ values, we can perform a hierarchical cluster analysis. Fig. 5.6 shows the dendrogram resulting from the classification. Four well-separated clusters are identified on this dendrogram, and we observe that the largest cities are clustered together and are characterized by a larger proportion of "random" flows (R) of individuals both living and working in parts of the city that are not the dominant residential and activity centers. This can be interpreted as an increased facility in bigger urban areas to commute from any part of the city to any other part.

5.1 Typology of origin–destination matrices

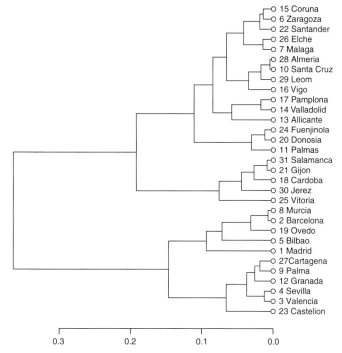

Figure 5.6. Dendrogram resulting from the hierarchical clustering of cities based on their ICDR values. The rank of a city in the hierarchy of population size is indicated in front of its name. The largest cities are clustered together. As cities get bigger, the "random" component (R) of their commuting flows increases, which signals that it is easier to commute from any place to any other in large cities. Figure taken from Louail et al. (2015).

In summary, the extraction of high-level information from the large OD matrices obtained with cell-phone data, allows to highlight several remarkable patterns in the data:

- Independently of the density threshold chosen to determine hotspots, the proportion of integrated flows (I) decreases with city size, while the proportion of random flows (R) increases.
- On average and for all cities considered here, individuals who live in residential main hubs and who work in employment main hubs (I flows) travel shorter distances than the others (C, D, R flows).
- When the city size increases, the largest impact is on convergent flows (C) of individuals living in smaller residential areas (typically in the suburbs) and commuting to important employment centers.
- The classification of cities based on the $ICDR$ values leads to groups with consistent population size, highlighting a clear relationship between the population size of cities and their commuting structure.

5.2 Modeling mobility patterns

As emphasized at the beginning of this chapter, understanding mobility patterns is a crucial issue for the science of cities. One of the oldest theoretical approaches is the gravity model (Erlander and Stewart 1990) and we briefly recall it here. We will then focus on modern approaches such as the radiation model and the closest opportunity model, which allows to explain the relation between income and commuting distance.

5.2.1 Statistics of flows: from gravity to radiation

Understanding the flows of individuals between two locations is a long-standing problem with a large number of applications, such as transport planning or epidemiology. From a quantitative point of view, Zipf (1946) proposed a simple form that can essentially be written as

$$T_{12} = K \frac{P_1 P_2}{d_{12}^\sigma}, \qquad (5.6)$$

where T_{12} is the number of individuals per unit time moving between locations 1 and 2 with populations P_1 and P_2. The flow should decrease with the distance d_{12} between these locations and enters this simple formula in a way similar to the usual gravity law in classical mechanics, but with an exponent σ that is usually different from two, that depends on the area considered, and that varies with time. This equation was subsequently modified and there is now a large number of variants; we refer the interested reader to the classical book by Erlander and Stewart (1990) for detailed discussions on this law. Recently, an improved version of this law was tested extensively against data by Lenormand et al. (2012), showing good agreement, at the cost of a larger number of parameters.

Several theoretical derivations were proposed for this law (Eq. 5.6) and in particular entropy maximization was proposed by Wilson (1969), and we briefly describe it here. As above, we denote by F_{ij} the number of trips from i to j and by F the total number of trips. For a given set $\{F_{ij}\}$ the number of ways to construct this configuration with F individuals is given by

$$\Omega = \frac{F!}{\prod_{ij} F_{ij}!}. \qquad (5.7)$$

In the absence of other information, Wilson postulated that the traffic is such that Ω is maximum subject to the following constraints (which is reminiscent of

5.2 Modeling mobility patterns

maximizing the entropy in a closed system)

$$\sum_j F_{ij} = F_i, \tag{5.8}$$

$$\sum_i F_{ij} = F_j, \tag{5.9}$$

$$\sum_{ij} F_{ij} C_{ij} = C, \tag{5.10}$$

where C_{ij} is the cost for one trip from i to j and the last constraint corresponds to the total amount of money C available for travel. The calculation of this maximum leads to the following result

$$F_{ij} = A_i B_j F_i F_j e^{-\beta C_{ij}}, \tag{5.11}$$

where A_i, B_j and β are constants ensuring that the constraints are met. In this expression the main dependence on space is in the exponential, where C_{ij} depends *a priori* on the distance between i and j. A severe drawback of this derivation is that we have to give an explicit form for this cost in order to compute F_{ij}: a power law in the gravity model for example implies a cost increasing logarithmically with distance, which is difficult to justify. In addition, it seems difficult to include in this approach effects such as congestion.

At this stage, there were no really satisfying approaches until Simini et al. (2012), who proposed a simple model – coined the "radiation model" (for its analogy with radiation processes in physics) – based on a small number of reasonable assumptions. In this radiation model, individuals are assumed to have a residence and the important ingredient is how they choose their job. The idea is that they assess a job according to some "quality" z which depends on a large number of factors such as income, working hours, conditions, etc. The larger this number, the higher are the expectations of this individual. Also, the larger the population where the agent looks for a job, the larger is this number, reflecting the fact that more populated areas propose a larger number of opportunities. For an individual living at location i with population m_i, a simple choice is to take

$$z(m_i) = \max\{z_1, z_2, \cdots, z_{m_i}\}, \tag{5.12}$$

where the z_i are independent random variables distributed according to some (unknown) distribution $p(z)$ (and with cumulative $P(z)$). This choice is indeed a simple way to construct a random variable that increases with m_i.

The surrounding locations offer jobs and if the location j has a population n_j, the proposed job benefit $\tilde{z}$ will be given by

$$\tilde{z}(n_j) = \max\{z_1, z_2, \cdots, z_{n_j}\}. \tag{5.13}$$

The larger the population, the larger this benefit will be. In the same spirit as radiation processes, Simini et al. (2012) proposed that an individual living in i will choose the closest job with benefit such that $\tilde{z}(n_j) > z(m_i)$. An individual starting from node i with population m_i will then work at node j with population m_j, with probability

$$P(i \to j | m_i, m_j, s_{ij}) = \int_0^\infty dz\, P_{m_i}(z) P_{s_{ij}}(<z) P_{m_j}(>z). \tag{5.14}$$

The first term in this integral is the probability to have $z(m_i) = z$, the second term is the probability that in the area between i and j all the s_{ij} offers are such that $\tilde{z} < z$, and the last term is the condition for accepting the job. More precisely, we have

$$P_{m_i}(z) = m_i P(z)^{m_i - 1} \frac{dP(z)}{dz}, \tag{5.15}$$

$$P_{s_{ij}}(<z) = P(z)^{s_{ij}}, \tag{5.16}$$

$$P_{m_j}(>z) = 1 - P(z)^{m_j}. \tag{5.17}$$

We thus obtain

$$P(i \to j | m_i, m_j, s_{ij}) = \int_0^\infty dz\, m_i P(z)^{m_i - 1} \frac{dP(z)}{dz} P(z)^{s_{ij}} (1 - P(z)^{m_j}), \tag{5.18}$$

and the change of variable $x = P(z)$ gives

$$P(i \to j | m_i, m_j, s_{ij}) = m_i \int_0^1 dx \left[x^{m_i + s_{ij} - 1} - x^{m_i + m_j + s_{ij} - 1} \right] \tag{5.19}$$

which finally leads to the result

$$P(i \to j | m_i, m_j, s_{ij}) = \frac{m_i m_j}{(m_i + s_{ij})(m_i + m_j + s_{ij})}. \tag{5.20}$$

This is the central result obtained by Simini et al. (2012) and shows that using the same distribution $p(z)$ for the expected and proposed job quality, the flow has the "universal" form Eq. (5.20). If we denote by $F_i = \sum_j F_{ij}$ the total flow of individuals "emitted" by i, the average flow between i and j is then given by

$$\langle F_{ij} \rangle = F_i \frac{m_i m_j}{(m_i + s_{ij})(m_i + m_j + s_{ij})}. \tag{5.21}$$

This last equation thus represents an alternative to the gravity law and has several nice features. The derivation is very clear and the assumptions well defined. Also, it is independent of $P(z)$ and has no free parameter.

An important difference between this form and the gravity law is how space enters the game. In this radiation model, space appears through the number of opportunities s_{ij} between i and j. In the case of a uniform population distribution of density ρ, we have $s_{ij} = s(r) = \rho \pi r^2$, where r is the distance between i and j, and the probability of finding a job scales for large distances r as

$$P(r)dr \sim \frac{2\pi r dr}{\pi \rho r^4}, \qquad (5.22)$$

leading to

$$P(r) \sim \frac{1}{r^3}. \qquad (5.23)$$

This result is interesting as it shows that the commuting distance can be broadly distributed, which will be confirmed empirically in the next section, and also theoretically within a framework that extends and generalizes the radiation model.

The comparison between data and this model has been discussed in the paper by Simini et al. (2012) and later on by other authors (see for example Lenormand et al. (2012)). In particular, improved versions of the gravity model can outperform this simple model for small geographical units (Lenormand et al. 2012). The main virtue of this radiation model is however its simplicity and the fact that it provides a simple framework for discussing a mechanism as complex as job searching. In this respect, it could probably be useful for further elaboration and theoretical discussion. In the next section, for example, we will see how a simple model which discusses the impact of income on the commuting distance, actually generalizes this radiation model.

These studies also echo the importance of the rank already highlighted in Liben-Nowell et al. (2005), where the probability that individuals at locations i and j are friends is written as

$$P[i \to j] \sim \frac{1}{\text{rank}(i, j)} \qquad (5.24)$$

where the rank is formally defined as

$$\text{rank}(i, j) = \#\text{nodes } w \text{ such that } d(i, w) < d(i, j) \qquad (5.25)$$

and is thus equal to the quantity s_{ij} defined above.

It is interesting to note here that in another study, Noulas et al. (2012) confirmed the importance of the rank in studying intra-urban movements. This convergence of various pieces of evidence supports the idea that the rank is a natural quantity in these problems.

5.2.2 Commuting and income

We saw in the previous section how we can model the distribution of commuting distances within a simple framework. Here we would like to go further in our understanding of commuting and relate it to socio-economical indicators. We will first describe the empirical results obtained for the commuting distance for Denmark, the UK, and the US. We will then present a framework (Carra et al. 2016) that generalizes the radiation model and allows to discuss the impact of income on commuting.

Empirical results

Carra et al. studied three datasets, for Denmark, the UK, and the United States. These datasets are produced by national agencies and national household surveys and record the commuting distance and the income range, for different years and the whole country, and mostly for all transportation modes. For the United Kingdom, the data is for years 2002–2012, for the United States three different years are available (1995, 2001, 2009), and for Denmark 10 years (2001–2010).

In Fig. 5.7, we see that the average commuting distance $\bar{r}$ increases with income (except for the US, where it is essentially flat), in agreement with standard models of urban economics (Alonso et al. (1964); Muth (1969); Brueckner et al. (2000)) where workers with higher incomes will have longer commuting distances. The data can reasonably be fitted by a power law of the form

$$\bar{r}(Y) \sim Y^\beta \qquad (5.26)$$

where Y is the income level and where the exponent β depends on the country considered:

- For the US, the fit gives an exponent $\beta \approx 0$, indicating that there is no clear trend.
- For the UK, the plateau occurs in the small income range $[10^2, 10^4]$ (GBP per year) around the commuting distance value $\bar{r} \approx 5$ miles. The fit on these data (for all modes and all years), gives an exponent value $\beta \approx 0.5$ (in the range $[0.53, 0.66]$ when considering different years) for an income larger than £5,000 per year.
- For the Danish data in contrast, we observe a strong dependence with a large exponent of order 0.8 for a yearly income larger than 250,000 DKK and smaller than 500,000 DKK (for smaller values of the income we observe a small plateau). Depending on the year considered, we find an exponent β that varies in the range $[0.61, 0.88]$.

More interesting is the full distribution of the commuting distance shown in Fig. 5.8. We observe that for all countries studied here, the distribution displays

5.2 Modeling mobility patterns

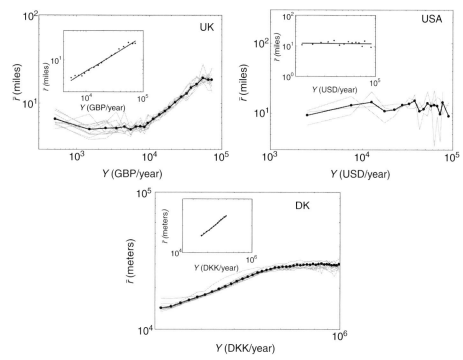

Figure 5.7. Average commuting distance versus income for different years. The gray curves correspond to a given year, and in black we show the commuting distance averaged over all years. (Top, left) UK data. This log–log plot displays a plateau for small values of income followed by a regime, when fitted by a power law (see inset), gives an exponent $\beta \approx 0.5$. In the inset the average commuting distance is averaged over all years and the power-law fit gives an exponent $\beta = 0.58$. (Top, right) US data. In this log–log plot we do not observe an income dependence. Indeed, a power-law fit gives an exponent $\beta \approx 0$. (Bottom) Danish data. The power-law fit on the commuting distance averaged over all years (in the inset) gives an exponent $\beta = 0.77$. Figures taken from Carra et al. (2016).

a slow decaying tail that can be fitted by a power law for large values of r

$$P(r) \sim r^{-\gamma} \qquad (5.27)$$

with $\gamma \in [2.7, 3.2]$ for Danish data, $\gamma \in [2.65, 3.0]$ for the UK data, and $\gamma \in [2.8, 3.3]$ for the US, depending on the income class considered.

There are two important facts that we can extract from these empirical observations. First, for all datasets studied here, the distribution of commuting distance is broad which implies that its variation range is extremely large. Indeed, we observe that with a non-negligible probability, individuals in Denmark, the UK, and the United States are commuting over distances of the order of a few

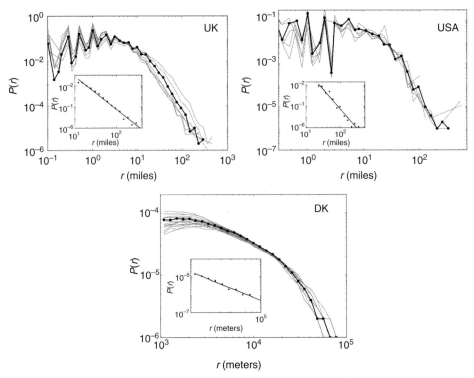

Figure 5.8. Commuting distance distribution for different income classes. The probability distribution is shown for different income classes (gray curves), and in black we show the distribution for a particular value of the income for which the fit on the tail is shown in the inset. (Top, left) UK data (averaged over all available years). The distribution displays a slow-decaying tail and a power-law fit on the tail ($r > 10$ miles) gives exponents in the range $[2.65, 3.0]$ (in the inset $\gamma = 3.16$). (Top, right) US data (averaged over all available years). The power-law fit on the tail ($r > 20$ miles) gives values for the exponent in the range $[2.8, 3.3]$ (in the inset $\gamma = 3.9$). (Bottom) Danish data (all years give the same result and we choose here to show the year 2008). The power-law fit on the tail ($r > 20$ km) gives values for β in the range $[2.7, 3.2]$ (in the inset $\gamma = 2.28$). Figures taken from Carra et al. (2016).

hundreds of kilometers. Second, the shape of the distribution and the large-distance behavior are remarkably similar among the different countries studied here, with a power-law decay with exponent of order $\gamma \approx 3$. It is interesting to note that this exponent is in agreement with predictions obtained from the radiation model (Simini et al. 2012) detailed in the previous section.

Theoretical modeling

The radiation model seems to be in agreement with the data described above, but the income variable does not enter this model and some further work is needed.

5.2 Modeling mobility patterns

The fundamental problems here are how individuals are actually choosing their job and how jobs are distributed in space. There have been many theoretical and empirical studies on this problem. The seminal contributions on search theory in economics (Lippman and McCall 1976; McCall 1970; Stigler 1961) rely on the central assumption that individuals choose among different job offers that arrive sequentially in time, by maximizing their expected net wage. This is however a very strong assumption, and ideally the model should be pushed to give testable predictions that can be checked against empirical data. In addition, and surprisingly enough, many standard models of job search (such as McCall 1970) do not integrate space and we will see here how to introduce it in the classical job search model of McCall. Despite adding space, the classical approach has difficulties in explaining the statistical regularities observed above and this is why we need to introduce the "closest opportunity model," Carra et al. (2016) which integrates income in this decision process and actually extends the radiation model.

The spatial optimal job search model. The stopping problem (Chow et al. 1971) is a classical example in optimal control and has been applied in many different areas (Wald 1947; Bradt et al. 1956; Shiryaev 1963; Haggstrom 1966; Rasmussen and Starr 1979). In particular, it has been applied in economics to the job search problem (Lippman and McCall 1976; McCall 1970; Stigler 1961) and we consider here as a starting point the McCall model (McCall 1970). In this model, a worker who is unemployed at time 0, reviews at every time step a random wage offer w drawn from a probability distribution f (with cumulative F). At each time step, the worker can either accept the current job offer and keep it forever, or she can pay a waiting cost c to discard the offer and wait for the next one ($-c$ can also be seen as the unemployment benefit). In this model, with an offer w at hand, the worker maximizes her expected value $v(w)$

$$v(w) = \left\langle \sum_{t=0}^{\infty} \mu^t y(t) \right\rangle, \tag{5.28}$$

where the brackets denote the average over the offer distribution and where $y(t)$ denotes the income of the worker at time t (the discount factor is $\mu < 1$). The worker's income is either $y(t) = w$ if she accepts the offer or $y(t) = -c$ if she refuses it. The classical way to solve this problem is to write the Bellman equation for this stopping process, which reads (Bellman 1957)

$$v(w) = \max \left\{ \frac{w}{1-\mu}, -c + \mu \int v(w') f(w') dw' \right\}. \tag{5.29}$$

The first term corresponds to the case where the worker accepts the offer, and the second term shows that if she refuses it, she incurs a cost c added to the average taken over all possible next offers w'. The optimal strategy that solves this equation is to accept the offer if it is larger than a *reservation wage* τ, and to refuse it if it is lower. From Bellman's equation Eq. (5.29), we obtain the following equation satisfied by this reservation wage

$$\frac{\tau}{1-\mu} = -c + \frac{\mu}{1-\mu}\left[\tau F(\tau) + \int_\tau^\infty w' f(w')dw'\right]. \tag{5.30}$$

By solving this equation, we obtain a function τ that depends on the offer distribution. The number N of trials before accepting a job offer thus follows a geometric distribution

$$P(N) = (1-p)^{N-1}p, \tag{5.31}$$

where p is the probability of accepting an offer

$$p = \int_\tau^\infty f(w)dw. \tag{5.32}$$

In order to introduce space in this model, we further assume that the worker reviews job offers in order of increasing distance from the home. We also assume that jobs are uniformly distributed in space with density ρ and that the probability of accepting an offer is still given by Eq. (5.32). If a worker has accepted the Nth offer, the probability that she has moved a distance r from her residence is given by a classical result for the Nth nearest neighbor in dimension $d = 2$ for uniformly distributed points (Diggle et al. 1983)

$$P(R=r|N) = \frac{2}{(N-1)!}\frac{1}{r}(\rho\pi r^2)^N e^{-\rho\pi r^2}. \tag{5.33}$$

The distribution of the commuting distance R is then given by

$$P(R=r) = \sum_{N\geq 1} P(r|N)P(N) \tag{5.34}$$

and since the distribution of N is geometric (Eq. (5.31)), we obtain

$$P(R=r) = 2p\rho\pi r e^{-p\rho\pi r^2}. \tag{5.35}$$

This distribution is not a power law and decreases exponentially over a scale of order

$$r_0 \sim \frac{1}{\sqrt{\rho p}} \tag{5.36}$$

where $1/\sqrt{\rho}$ corresponds to a typical interdistance between different offers. We also note that the average commuting distance decreases if the spatial density of opportunities ρ increases. A decrease in the number of job openings during economic downturns then leads to increasing commuting distances.

The important fact is that the result Eq. (5.35) is not consistent with the empirical evidence and the optimal strategy seems to be an assumption not backed up by data. At first glance this is however not unexpected. Indeed, as noted above, the empirical observations reveal that the commuting distance varies over a very broad range, which suggests that the individual's behavior might not be optimal and might depend on a large number of uncontrolled factors. We thus have to find an alternative model for explaining the empirical findings. In the next section, we discuss such a model and compare its predictions with data.

The closest opportunity model. In this model, we consider that the job search process is governed by the two following ingredients:

- workers search through space and not through time,
- jobs are chosen based on some "quality" aspect and not on the wage (see for instance Hornstein et al. 2007; Hall and Mueller 2013).

We assume that each worker is characterized by a skill level I that we will identify with the income. The density of jobs $\rho(I)$ relevant for the worker depends on the skill level and we assume that it is simply given by

$$\rho(I) = \frac{\rho_0}{I^\alpha}, \qquad (5.37)$$

such that higher-skilled jobs are less dense than lower-skilled jobs. For each skill level, we assume that the spatial distribution is uniform (with density $\rho(I)$). The exponent α depends on the country under consideration and reflects many exogenous factors concerning job offers at a certain skill level (Hornstein et al. 2007; Hall and Mueller 2013).

As stated above (and in the same spirit as the radiation model), we assume that each job is characterized by a random "quality" X that encodes many factors. The job quality is distributed according to f (with corresponding cumulative distribution F), and job qualities are independent. We assume that a given worker has a *reservation quality* value τ (in the same spirit as the reservation wage), and she will keep expanding her search radius until this threshold is met. We denote by R the commuting distance and its cumulative (for a given reservation quality) thus reads

$$P(R \le r | \tau) = P(X_{[0,r]} \ge \tau) = 1 - F(\tau)^{\rho \pi r^2}. \qquad (5.38)$$

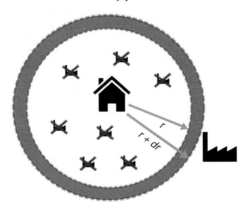

Figure 5.9. Illustration of the argument leading to Eq. (5.40). Figures taken from Carra et al. (2016).

We now take into account that workers have different search costs and different expectations, which leads them to have different reservation qualities. We thus naturally consider that the reservation quality is a random variable distributed according to a density $g(\tau)$. The cumulative distribution of commuting distances is then

$$P(R \leq r) = \int g(\tau) P(R \leq r | \tau) d\tau, \qquad (5.39)$$

with corresponding density

$$P(R = r) = -2\rho\pi r \int g(\tau) F(\tau)^{\rho\pi r^2} \log F(\tau) d\tau. \qquad (5.40)$$

The first term in this integral is the probability that a worker has reservation quality τ, the second term is the probability that all offers are below τ in the disk of radius r, and the last term (the logarithm) corresponds to the probability that at least one offer is above τ in the circular band $[r, r + dr]$ (indeed $1 - F^{2\pi\rho r dr} \approx -2\pi\rho r dr \log F$). We show in Fig. 5.9 a simple illustration of this process.

A simple and natural assumption for the distribution of the reservation quality τ is to choose the same as the distribution of job quality. Then Eq. (5.40) simplifies in a remarkable way as follows

$$P(R = r) = -2\rho\pi r \int f(\tau) F(\tau)^{\rho\pi r^2} \log(F(\tau)) d\tau$$

$$= -2\rho\pi r \int_0^1 x^{\rho\pi r^2} \log x \, dx$$

$$= \frac{2\rho\pi r}{(1 + \rho\pi r^2)^2}. \qquad (5.41)$$

Under these assumptions, the distribution of commuting distances does not depend on the distribution of job quality, an effect that was already observed in the specific case of the radiation model (see Section 5.2 and Simini et al. 2012), and the model proposed here can then be considered as a microfoundation for this type of process. This also means that we may generalize the interpretation of the model: we may allow the distribution of job quality to be specific to each worker, since this has no consequence for the distribution of commuting distances.

In contrast to the job-search models of the previous section that displayed a rapid exponentially decaying tail, we observe here that the distribution is slowly decaying as $P(R=r) \sim r^{-3}$ for large r. The average commuting distance is easily computed here

$$\bar{r} = \frac{1}{2}\sqrt{\frac{\pi}{\rho}} \tag{5.42}$$

and, replacing ρ by ρ_0/I^α, the average commuting distance reads

$$\bar{r}(I) = \frac{1}{2}\sqrt{\frac{\pi}{\rho_0}} I^{\alpha/2}, \tag{5.43}$$

which is a power law with exponent $\beta = \alpha/2$. We thus observe a behavior $\bar{r} \sim I^{\alpha/2}$ for the closest opportunity model, where α depends on the country considered, and we can interpret the empirical results in this framework. For the US, we observe an exponent $\beta_{US} \approx 0$, indicating that the density of jobs seems to be independent of the skill level in this country. For the UK and Danish datasets, we observe a non-zero exponent with $\beta_{UK} \approx 1/2$ for the UK and a larger value for Denmark, $\beta_{DK} \approx 0.8$. These results indicate that the larger the skill level (or income in this framework), the smaller is the density of corresponding jobs, a result that is even more marked in Denmark than in the UK.

When we express ρ as a function of I, the distribution of commuting distances conditional on income becomes

$$P(R=r|I) = \frac{2\rho_0 \pi r I^\alpha}{\left(I^\alpha + \rho_0 \pi r^2\right)^2}. \tag{5.44}$$

This result Eq. (5.44) implies a simple scaling that can be checked empirically. Indeed, if we rescale the commuting distance by $I^{\alpha/2}$, $u = r/\sqrt{I^\alpha}$ (and the distribution according to the corresponding factor), all the curves for different incomes should collapse onto the unique curve given by

$$P(u) = \frac{2\pi \rho_0 u}{(1 + \rho_0 \pi u^2)^2}. \tag{5.45}$$

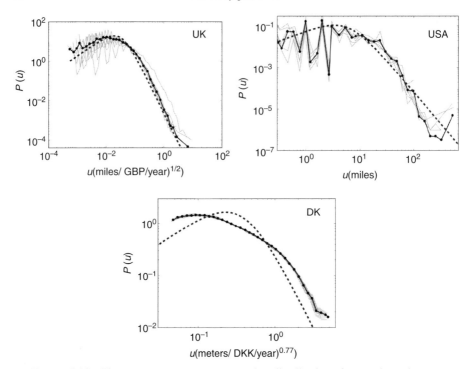

Figure 5.10. The gray curves represent the distribution for a given income class and the black line the average over all income classes. The dotted line is the one-parameter fit (for the averaged rescaled distribution) with the function predicted by the opportunistic model, Eq. ((5.45)). (Top,left) UK data (averaged over available years). The collapse is obtained with $\beta = 0.5$. (Top, right) US data (averaged over available years). (Bottom) DK data (year 2008). Collapse of the rescaled probability with $\beta = 0.77$. Figures taken from Carra et al. (2016).

In Fig. 5.10 we plot the rescaled commuting distance distribution for different income categories and we observe a very good collapse, except for the lower income category in the UK for which the square root behavior is not applicable. The agreement of Eq. (5.44) with data is very good for the UK and the United States, but there are some discrepancies in the Danish case even if the collapse confirms the scaling form Eq. (5.45). It seems that the model reproduces relatively well the large-distance behavior, but underestimates the heterogeneity of individual behavior in this case.

In summary, we have seen here that empirical data do not support the optimal strategy model for the job-search process. Instead, a simple model, based on the closest opportunity that meets the expectation of the individual, is able to predict correctly the behavior of the average commuting distance with income in terms of the density of job offers. This model is also able to provide the correct "universal" form of the distance distribution and its broad tail. Although further studies on more countries are certainly needed, this stochastic model provides a microscopic foundation for a large class of mobility models and opens many interesting directions in modeling commuting patterns and leading to testable predictions. After having considered commuting patterns we will study in the next section the distribution of trips for any purpose.

5.3 Human mobility: Levy flights or accelerated walkers?

The recent availability of data allowing monitoring the position of individuals triggered a wealth of quantitative studies on human mobility. In particular, displacements are often described by a Lévy type of walk, characterized by many small movements and some rare long jumps Δr. More precisely, when the distribution of the lengths $P(\Delta r)$ is a power law, the jumps are called *Lévy flights* (see for example Zaburdaev et al. 2015).

The importance of $P(\Delta r)$ comes from its central role in modeling human traveling behavior, and many authors studied this quantity, without reaching however a consensus.

Brockmann et al. (2006), using modern data on human traveling statistics, invoked a scale-free *continuous time random walk* in order to explain the movement of dollar bills. In this study, a dollar note trajectory is reconstructed from reports at various locations. This random sampling however is a strong limitation and cast some serious doubts about the validity of this study. Many studies followed and added to the confusion: at a large scale (national or inter-urban), a long tail behavior has been observed (Brockmann et al. 2006; Gonzalez et al. 2008; Song et al. 2010a; Cheng et al. 2011; Liu et al. 2012; Noulas et al. 2012; Yan et al. 2013; Hawelka et al. 2014), characterized by a power-law decay for long displacements. At a smaller (urban) scale, the distribution seems to be exponentially decreasing (Bazzani et al. 2010; Gallotti et al. 2012; Liang et al. 2012; Kang et al. 2012; Wang et al. 2015; Liu et al. 2015).

Most of the fits can be described as particular cases of the function

$$P(\Delta r) = \frac{1}{(\Delta r + \Delta r_0)^\beta} e^{-\Delta r/\kappa} \tag{5.46}$$

with different values for the parameters Δr_0, β, and κ. We note that further studies proposed a polynomial form close to an exponential behavior for private cars (Bazzani et al. 2010), two different behaviors for urban and inter-urban trajectories for cars and taxis (Gallotti et al. 2012; Liu et al. 2015), or a lognormal distribution for individual GPS tracks (Zhao et al. 2015). We report here the various results in the Table 5.1.

Table 5.1. *Parameter values for the fit of the displacement distribution with a truncated power law found in previous studies. This list includes studies on different data sources and spatial or temporal scales. Only fits consistent with the function defined in (Eq. 5.46) are presented here. The case $\kappa = \infty$ is associated with non-truncated power laws, while $\beta = 0$ to exponential distributions. $\Delta r_0 = 0$ correspond to cases where this parameter was omitted in the fit, while $\beta = 0$ when this value is not defined.*

Data Source	Trajectories	β	κ (kms)	Δr_0 (kms)
Dollar Bills [*]	464K	1.59	∞	0
Cell Phones [†]	100K	1.75	400	1.5
Cell Phones [†]	206	1.75	80	1.5
Cell Phones [‡]	3M	1.55	100	0
Cell Phones [+]	7K	0	[2, 5.8]	-
GPS tracks [♭]	101	[1.16,1.82]	∞	0
Location Sharing [♮]	220K	1.88	∞	0
Location Sharing [**]	900K	1.50	∞	2.87
Location Sharing [**]	900K	4.67	∞	18.42
Location Sharing [††]	521K	0	300	-
Taxis [++]	12K	0	4.29	-
Taxis [#]	7K	1.2	10	0.31
Taxis [##]	30K	0	[2, 4.6]	-
Taxis [◦]	1100	[0.50,1.17]	[4.5, 6.5]	0
Travel Diaries [†††]	230	1.05	50	0
Tweets [###]	13M	1.62	∞	0

[*]: Brockmann et al. (2006); [†]: Gonzalez et al. (2008); [**] Noulas et al. (2012); [‡]: Song et al. (2010a); [+]:Kang et al. (2012); [♭]: Rhee et al. (2011); [♮]: Cheng et al. (2011); [††]: Liu et al. (2014); [++]: Liang et al. (2012); [#]: Liu et al. (2012); [##]: Wang et al. (2015); [◦]: Tang et al. (2015); [†††]: Yan et al. (2013); [###]: Hawelka et al. (2014). Table from Gallotti et al. (2016a).

An important factor that causes the absence of consensus is the nature of these studies. Most authors are performing a fit but usually without a clear justification of the function used. In some studies, the authors then tried to explain the fit obtained with some mechanism (such as Song et al. 2010b). This is however a process that is upside-down, and instead, a theoretical model should be constructed in order to provide a guide allowing the testing of simple assumptions about the mechanisms governing mobility.

5.3.1 Back to basics: empirical observations

In order to characterize a trip, we need to distinguish pauses from actual traveling. Both phases are characterized by their duration, and in the traveling phase, the

5.3 Human mobility: Levy flights or accelerated walkers?

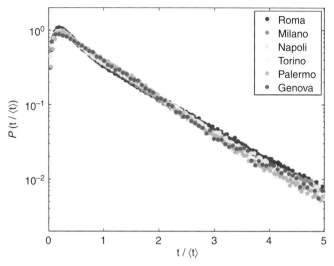

Figure 5.11. Travel time in the 6 largest Italian cities. The probability distribution is an exponential function and the difference between cities is therefore fully encoded in a single parameter: the average travel-time $\langle t \rangle$. Figure from Gallotti et al. (2016).

velocity is the crucial parameter. These two steps need to be distinguished for modeling trips, since costs are in general associated with trips, while a positive utility can be associated with the activity performed during stops (Axhausen and Gärling 1992). With GPS data, these two phases are easy to separate as the transition is identified exactly by the moment when the engine is turned on or off. This allows an accurate evaluation of the probability distributions of travel times t and pause times τ. This is however an important limitation for other sources of data such as those obtained by calls or social networking, which capture the spatial character of individuals' movements (Lenormand et al. 2014), but are necessarily coupled to the bursty character of human communication (Barabasi 2005) and thus are not suitable for an exhaustive temporal description of human mobility.

From GPS data, we observe that the distribution of cars' travel-times is characterized by an exponential tail (Fig. 5.11). For public transport systems, we also observe a rapidly decreasing tail for the travel-times between metro stations in London, evaluated from the *Oyster Card* ticketing system (Roth et al. 2011). Travel times for cars however depend on the city population (Louf and Barthelemy 2014a): the larger the city, the worse are the congestion effects. This is confirmed by the results of Gallotti et al. 2015, which show that the average travel-time $\langle t \rangle$ in different cities falls in the range [9, 18] minutes, with a very slow growth correlated with population $\langle t \rangle \propto P^{\mu}$, where $\mu = 0.07 \pm 0.02$. As a consequence of this, the average travel time differs from one city to another, but if we rescale the time by its average, we observe in Fig. 5.11 that the distribution $P(t/\langle t \rangle)$ appears to be universal across different cities and is given by

$$P(t) = \frac{1}{\langle t \rangle} e^{-t/\langle t \rangle}. \tag{5.47}$$

The average value $\langle t \rangle$ contains then all the information needed for describing cars' travel times in a particular city.

The distribution of pausing times $P(\tau)$ does not display remarkable differences between cities for stops shorter than a day. Similarly to waiting times of email communications (Barabasi 2005), this distribution is close to a power law

$$P(\tau) \sim \tau^{-\gamma} \tag{5.48}$$

with an exponent $\gamma \approx 1$ (Gallotti et al. 2012).

The distributions $P(t)$ and $P(\tau)$ describe statistical properties of mobility patterns in time and in order to understand how these quantities translate into displacements in space, we need information about travel speeds. Obviously, speeds are not constant but depend on many factors such as transportation modes, congestion, etc. For private cars, longer trips tend to be faster since they can make better use of the road infrastructure. The same is true for public transport, where speed increases with the trip's length at an urban level owing to the use of faster transportation modes (Gallotti and Barthelemy 2014).

The speed depends on the trip duration and we present its distribution for different time interval bins. For trips of duration t between 10 minutes and an hour, the speed distribution $P(v)$ has an exponential tail. For longer times $t > 1$ h, the tail becomes shorter as a natural consequence of the speed limit (which is 130 km/h on Italian highways). More interestingly, the average speed value is enough to describe completely the dependence on t, and we observe from the data that it displays a linear growth

$$\langle v(t) \rangle = v_0 + at \tag{5.49}$$

with a uniform acceleration $a = 16.7$ km/h^2 and a "base speed" $v_0 = 17.9$ km/h (Fig 5.12(left)). This trend is observed for $t < 2$ h, followed by a saturation of $\langle v(t) \rangle$ towards the limiting average speed of ≈ 60 km/h. We note that for single cities there are of course local differences and deviations from the linear fit that are probably due to the inhomogeneity of the street infrastructure. For public transport, we observe a similar linear growth for the average velocities of trajectories in the public transport system (Fig 5.12(right)). In this case we do not have actual individual trajectories, but if we assume that there is a uniform travel demand and that travelers make the shortest time-respecting path between origin

5.3 Human mobility: Levy flights or accelerated walkers?

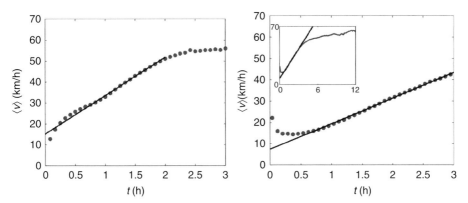

Figure 5.12. The average acceleration of trips with private cars and public transportation. (Left) The average $\langle v \rangle$ grows linearly with the trip duration t for trips shorter than 2 hours, then saturates to a value ≈ 60 km/h. The solid line represents the linear fit with $\langle v(t) \rangle = v_0 + at$ with an initial speed $v_0 = 17.9$ km/h and an acceleration $a = 16.7$ km/h². (Right) We observe also for public transport trajectories (linking all possible origins and destinations in Great Britain) a uniform acceleration for trips of duration between 30 minutes and 3.5 hours. The solid line represents the fit with $v_0 = 7.4 \pm 0.4$ km/h and $a = 12.0 \pm 0.2$ km/h². Figures taken from Gallotti et al. (2016a).

and destination, we can estimate velocities (Gallotti and Barthelemy 2014). Short trips, of less than 30 minutes, tend to be faster thanks to the likely absence of time-consuming connections. For $t \in [0.5, 3.5]$ hours, the growth is again linear but both base speed $v_0 = 7.4 \pm 0.4$ km/h and acceleration $a = 12.0 \pm 0.2$ km/h² are smaller than in the case of private transport. In addition, we observe that for public transport the distribution $P(v)$ is Gaussian-like. A possible explanation for the difference between the exponential shape of $P(v)$ for private transport and the Gaussian-like distribution for public transport might lie in the fact that in public transport the average speed is the weighted average of velocities associated with each edge of the transportation network (plus the effect of waiting time), leading to a normal distribution according to the central limit theorem.

5.3.2 Modeling the hierarchy of modes

From the data, we thus observe that the trip duration distribution has an exponential tail and that the average speed increases with the trip duration. This effective acceleration is a crucial ingredient for explaining the shape of $P(\Delta r)$ at all scales. Indeed, the probability distribution of the cars' displacements results from the combination of the trip duration distribution and the speed distribution $P(v)$, whose average depends itself on the trip duration.

The effective acceleration that we observe empirically in both private and public transportation is necessarily related to the hierarchical nature of transportation networks. Indeed, it is likely that faster transportation modes or faster roads are used more frequently for longer trajectories (Gallotti et al. 2012; Gallotti and Barthelemy 2014). In order to test this idea, it is then natural to propose a simple stochastic model based on the hierarchical structure of transportation.

In this simplified hierarchical model, we assume that the transportation network has n layers L_n, corresponding to different travel speeds v_n (see the schematic representation in Fig. 5.13). For the sake of simplicity, we assume that the velocity differences between layers are constant and equal to δv. An individual starts his trip of duration t in L_0, with base speed v_0, while on layer L_k he is traveling faster, at speed $v_k = v_0 + k\delta v$. The acceleration kicks from one layer to another happen at random times Poisson-distributed. In the simple case where we have two layers, we can solve this problem analytically. Indeed, the probability of a jump at time t^* is given by

$$P(t^* = t) = p e^{-pt}, \qquad (5.50)$$

and the position at time t of the traveler is

$$x(t) = v_0 t \theta(t^* - t) + \theta(t - t^*)\left[v_0 t^* + v_1(t - t^*)\right], \qquad (5.51)$$

where $\theta(x)$ is the Heaviside function. By averaging the position over t^*, we obtain for the average speed $\bar{v} = \bar{x}/t$ (the bar $\bar{\cdot}$ denotes the average over different trips)

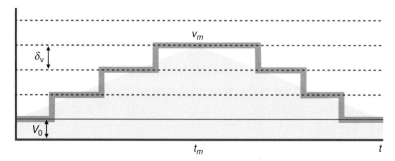

Figure 5.13. Schematic representation of the model. The trajectory starts with an acceleration phase, where the velocity is increased by constant kicks equal to δv. These kicks happen at random times with a uniform probability per unit time p. When approaching the destination, there is a deceleration phase with kicks of $-\delta v$ and the same probability p. Given the symmetry of the problem, the maximal speed v_m is reached on average at $t_m = t/2$. This maximal speed depends on δv and p: $v_m = v_0 + k(t_m p)\delta v$, where $k(\lambda)$ follows a Poisson distribution of average λ. The average speed $\bar{v}$ of the trajectory can be evaluated by approximating the step function, which defines the shaded area, as $\bar{v} = (v_0 + v_m)/2$. Figure taken from Gallotti et al. (2016a).

the following expression:

$$\bar{v}(t) = v_1 + (v_0 - v_1)\frac{1 - e^{-pt}}{pt}. \qquad (5.52)$$

We can examine this expression for the different limits:

- $pt \ll 1$: the average speed grows linearly from the base value v_0

$$\bar{v}(pt \ll 1) \approx v_0 + \frac{1}{2}pt(v_1 - v_0) \qquad (5.53)$$

- for $pt \gg 1$ the average speed converges asymptotically to v_1:

$$\bar{v}(pt \gg 1) \approx v_1 - (v_1 - v_0)/pt \qquad (5.54)$$

This simple model thus recovers, in some regime the linear growth of speed with the duration of the trip, and also the tendency to reach a limiting speed observed in simulations (Gallotti et al. 2016a). In particular, the uniform acceleration is naturally proportional to $p\delta v$.

More generally, we can estimate the distribution of velocities. For this, we assume that a trip is composed of two phases (see Fig. 5.13) of approximately the same duration. In the first one, the trajectory progressively jumps from slower to faster transportation modes. In the second phase, there is the inverse process where the trajectory progressively jumps down the layers until it finally reaches the base layer at time t. We also assume that the process has a Poissonian character, where all individuals have the same probability per unit time p of jumping to the successive layer and to change their velocity (both in the ascending and descending phases). Using these assumptions, we can then:

- estimate the maximal speed $v_m = v_0 + k(t_m)\delta v$, where $k(t_m)$ is the number of jumps at mid trajectory $t_m = t/2$;
- approximate the average speed as

$$\bar{v} = 1/t \int_0^t v(t)dv = (v_0 + v_m)/2. \qquad (5.55)$$

Since the process is Poissonian, $\langle k(t_m)\rangle = pt_m$ and the average speed is given by

$$\langle \bar{v}(t)\rangle = v_0 + \frac{p\delta v}{4}t \qquad (5.56)$$

where the brackets denote the average over the Poisson variable k. The average speed thus grows linearly with t before reaching the saturation imposed by the finite number of layers. Remarkably enough, this simple model allows us also to predict the shape of the conditional probability distribution $P(\bar{v}|t)$. Indeed, the number of jumps k is distributed following the Poisson distribution $P(k) = \frac{e^{-\lambda}\lambda^k}{k!}$

with $\lambda = pt/2$. Using the Gamma function as the natural analytic continuation of the factorial $k! = \Gamma(1+k)$, we obtain the distribution

$$P(\bar{v}|t) = \frac{1}{\delta v'} \frac{\exp\left(-p't + \frac{\bar{v}-v_0}{\delta v'}\log(p't)\right)}{\Gamma(1+\frac{\bar{v}-v_0}{\delta v'})} \quad (5.57)$$

where $p' = p/2$ and $\delta v' = \delta v/2$. The empirical distribution of the velocities at fixed time $P(\bar{v}|t)$ for travel times ranging between 5 and 180 minutes is consistent with Eq. (5.57) with $\bar{t} = 0.30$h, $\delta v' = 20.9$ Km/h, $p' = 1.06$ jumps/h and $v_0 = 17.9$ km/h. This makes the velocity gap $\delta v \approx 40$ km/h, remarkably consistent with the progression of the most common speed limits in Italy: 50 km/h (urban), 90 km/h (extra-urban), 130 km/h (highways). This result suggests that a multilayer, hierarchical transportation infrastructure can explain the constant acceleration observed in both public and private transportation. This model also allows us to estimate the base speed v_0 as the intercept value in Fig. 5.12, and the acceleration a is expected to be proportional to the probability of a jump to faster layers p and to the gap between layers δv.

Finally, the shape of the displacement distribution $P(\Delta r)$ can be computed as a superimposition of Poisson distributions

$$P(\Delta r) = \int_0^\infty dt \int_{v_1}^{v_2} d\bar{v}\, \delta(\Delta r - \bar{v}t) P(t) P(\bar{v}|t)$$

$$= \int_0^\infty \frac{dt}{\bar{t}} \frac{\exp\left(-\left(p'+\frac{1}{\bar{t}}\right)t + \log(p't)\frac{\Delta r/t - v_0}{\delta v'}\right)}{\Delta v' \Gamma\left(1+\frac{\Delta r/t - v_0}{\delta v'}\right)} \quad (5.58)$$

where δ is the Dirac delta function. The exact form for Eq. (5.58) cannot be analytically computed but the limiting behavior of for large Δr is of the form

$$P(\Delta r) \sim \frac{1}{\Delta r^\gamma} e^{-C\Delta r^\delta} \quad (5.59)$$

with $\gamma = 1/4$ and $\delta = 1/2$. In particular, this distribution is not broad, in clear contrast to Lévy flights, which have divergent moments and are governed by large fluctuations. Since the distribution $P(\Delta r)$ is not a fat tail distribution, all the phenomena associated with Lévy flights, such as super-diffusion for example, are not expected for randomly accelerated walks. In Fig. 5.14, we show this prediction for $P(\Delta r)$ with $v_0 = 17.9$ km/h, estimated as the intercept of the fit in Fig. 5.12; $p' = 1.06$ jumps/h $\delta v = 20.9$ km/h, estimated from the fit of $P(\bar{v}|t)$; and $\langle t \rangle = 0.30$ h, coming from the average of the travel-times for all selected trips. We note the good agreement of this prediction with the empirical data, and we stress that the

5.3 Human mobility: Levy flights or accelerated walkers?

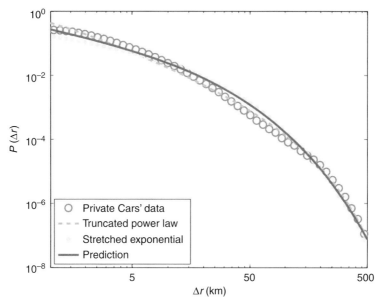

Figure 5.14. Distribution of displacements. We show (circles) the aggregated empirical distribution $P(\Delta r)$ for all the $\approx 780,000$ cars in the GPS dataset studied in Gallotti et al. (2016a). The dashed line represents a fit with a truncated power law of the form $P(\Delta r) \propto (\Delta r + \Delta r_0)^{-\beta} \exp(-\Delta r/\kappa)$ with $\beta = 1.67, \kappa = 84.1$ km, $\Delta r_0 = 0.63$ km. The coefficients of the fit are close to those found for cell-phone data in the United States (Gonzalez et al. (2008)). The dots represents the fit with the stretched exponential associated to uncorrelated accelerations model. The solid darker line shows the prediction based on the hierarchical model Eq. (5.58). This prediction is of remarkable quality that can reasonably compared with the commonly used direct fit with a truncated power law with 3 free parameters. Figure taken from Gallotti et al. (2016a).

curve proposed here is not an a posteriori best fit of the empirical $P(\Delta r)$ but the curve predicted from Eq. (5.58), knowing the average travel-time $\bar{t}$ and with the values v_0, p', $\delta v'$ used in the description of the empirical $P(\bar{v}|t)$.

The fact that we can fit the data by different forms, as the (truncated) power-law fit proposed in various studies (Brockmann et al. 2006; Gonzalez et al. 2008; see Table 5.1) or a stretched exponential, is an illustration of the difficulty in extracting mechanistic information from empirical data using only macroscopic statistical laws, without taking into account the dynamical properties of the underlying processes. In particular, urban mobility seems not to be described by a Lévy process as inferred from truncated power-law fits, but its random nature is governed by transitions between modes or roads with different velocities. In contrast with purely empirical works that led to Lévy flight models, a simple model based on the interaction between individuals and transportation networks is able to predict the displacement distribution, and to provide a possible explanation of the processes generating such macroscopic patterns. This type of model can be easily extended by integrating other processes and more information about the structure of transportation networks, for example. In this respect this approach could serve as an interesting starting point for more sophisticated models and for testing the relevance of various ingredients.

6
Multimodality in cities

In the previous chapter, we discussed the analysis and modeling of mobility patterns in cities. However, as cities expand, their transportation networks are also growing, with increasing interconnections between different transportation modes. In large cities, we can now choose the transportation mode to travel from one point to another, and this trip can even involve several different modes. This multimodality is a new aspect of large cities and brings new questions and problems. From the users, point of view, it becomes difficult to deal with the huge amount of information needed for describing the different transportation networks and their interconnections. From the transport agencies point of view, the managing task becomes harder because the different modes are usually run by separate agencies; this renders optimization difficult owing to the large number of aspects that have to be taken into account (Guo and Wilson 2011).

In particular, an important problem concerning multimodality is the synchronization between different modes. For example, on average in the UK, 23% of travel time is lost in connections for trips with more than one mode (Gallotti and Barthelemy, 2014). This lack of synchronization between modes induces differences between the theoretical quickest trip and the "time-respecting" path, which takes into account waiting times at interconnection nodes.

In order to address these problems and more generally to understand the impact of the coupling between modes, we need new tools in order to identify the main factors that govern their efficiency. The multilayer network approach seems to be the most convenient framework for studying these systems (Kivelä et al. 2014; Boccaletti et al. 2014). In this framework, each layer represents a mode and intermodal connections are represented by inter-layer links. In this chapter we discuss some aspects of multimodality and present tools for measuring and characterizing these coupled networks and their efficiency as a whole.

6.1 A multilayer network view of urban navigation

6.1.1 Empirical observations of multimodality

We first describe empirical results obtained by Gallotti and Barthelemy (2014) from timetables for the whole UK and for all transport modes. We note that these results were not obtained for traffic data (that are usually difficult to get). Instead we used these timetable data and the assumption of uniform demand. This allows us to test structural properties of these systems, but in order to get more precise results, one should use real OD matrices. At the national level (Fig. 6.1), the vast majority of short trips are made within the bus layer, and the rail system becomes dominant for inter-urban trips of length more than approximately 40 kms. Air transportation emerges naturally for longer distances above 200 kms, and its importance increases significantly for distances of order 400-500 kms (e.g. Glasgow–Birmingham or Glasgow–London Luton), finally becoming dominant for trips longer than 700 kms, connecting for example the southern part of England with the northern part of Scotland.

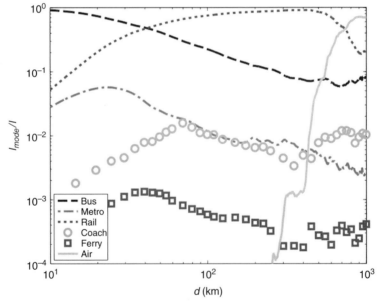

Figure 6.1. Fraction of distance covered by the different modes for trips in the whole of Great Britain. Short trips are mostly done by bus. Rail becomes then dominant at 40 kms and air travel is dominant for trips of distance of order 700 kms. Other transportation modes play a secondary role, with peaks at 22kms for the metro, 40 kms for ferries and 70 kms for coaches (which are increasingly used for long distances as a means to connect to airports). Figure taken from Gallotti and Barthelemy (2014).

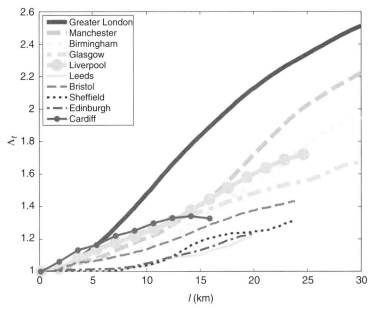

Figure 6.2. Multimodality in urban areas. The number of modes Λ_t used in trips through urban areas versus the trip length ℓ. Figure taken from Gallotti and Barthelemy (2014).

At the urban level, transportation modes that capture a significant fraction of the time-respecting paths are buses, railroads and, when available, the metro and tramway layers. Only large cities can afford significant rail-based elements (trains, metro, tram) in their public transport systems and therefore can have a high propensity for interchange (see Fig. 6.2). Indeed, we observe that larger cities (London, Manchester, Birmingham) show a particularly marked trend towards multimodality, with an average of more than 2 modes for trips longer than 20 kms.

Other transportation modes, such as ferries and coaches, play a secondary role at an urban level (and air transportation is naturally out of the game at this scale). If cities have enough suitable street space dedicated to buses and Bus Rapid Transit systems, they are even able to outperform metro and rail systems (Daganzo 2010). Depending on the length of the trip, the public transport system offers different optimal solutions. In the example of the UK shown in Fig. 6.3, different strategies emerge at different spatial scales. The use of fast transportation modes emerges progressively with increasing ℓ, with a higher rate in larger cities such as London, Manchester and Birmingham, where metro and tramway systems are present. As subways have high frequency, are fast and not affected by congestion, they are naturally used as a quicker alternative to buses across city centers. Nevertheless, owing to its limited accessibility, the largest fraction of short trips are done in the

164   *Multimodality in cities*

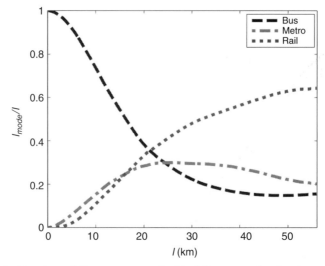

Figure 6.3. Fraction of distance covered with different modes for time-respecting paths through London. The bus system is covering most of the short trips, whereas the advantage of using the metro and rail systems emerges progressively for longer distances. Metro networks were naturally developed for answering urban transportation demand and we see that their use competes with the rail system for distances shorter than 15 kms (for larger distances, rail prevails). Figure taken from Gallotti and Barthelemy (2014).

bus layer, also in cities where the system has an extended network of metro lines. The metro layer is in competition with the rail layer, which has higher speed but lower departure frequencies. In cities with high multimodality (i.e. high average number of modes used per trip), the rail network attracts the largest part of the mobility at distances much lower than on the national scale. Indeed, for London and Manchester, the distance traveled by train exceeds that by buses at $\ell \approx 20$ and $\ell \approx 30$ km.

In order to identify the role of the temporal and multilayer aspects of the network in the structure of the time-respecting paths, we can detail how the total travel time can be decomposed into different components:

- The riding time. In order to take into account the multimodal aspect of the network, we discriminate riding times per layer.
- Waiting and walking times at interchanges. As for riding times, we separate intra-layer from inter-layer waiting times.

This wide spectrum of temporal quantities forms what we can call the "anatomy" of a trip, and is represented in Fig. 6.4 for the city of London. It has been observed (Gallotti and Barthelemy 2014) that in all cities but London, waiting times are longer than riding times for short trips. These waiting times are mostly intra-layer waiting due to bus–bus interchanges. If the city network is particularly

6.1 A multilayer network view of urban navigation

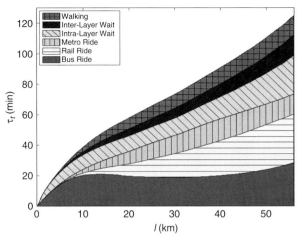

Figure 6.4. The anatomy of the transportation networks in London. The average structure of the total travel time changes with the trip length ℓ. Waiting times are mostly due to bus–bus interchanges. In Greater London, for trips longer than 20 kms, most of the riding time is spent on the metro or the rail. Figure taken from Gallotti and Barthelemy (2014).

multi-modal, inter-layer waiting and walking times start playing a significant role for longer values of ℓ, but only in London do they exceed intra-layer waiting times. As mentioned in the introduction to this chapter, for all UK cities the time spent in connections (walking and inter-layer waiting times) represents on average a significant fraction ($23 \pm 6\%$) of the total travel time. Short trips are dominated by intra-layer waiting times, long trips by riding times. In cases where the multimodality becomes dominant, inter-layer waiting and walking times, together with the fast layers' cruise speed, become instead the most relevant quantities for optimizing the system.

6.1.2 Characterizing the multilayer system

Multilayers systems made of different networks connected by inter-layer links cannot be considered as "large networks." The reason for that is the heterogeneity of links and the different dynamics on each layer that prohibit the identification of this system with a large, homogeneous complex network. As such, the standard tools for characterizing large networks are not useful and we have to construct specific tools for multilayer systems. Here we discuss some of these tools that are adapted to the specific case of transportation systems.

Interdependency

In many studies on multilayer networks, the word *coupling* describes the topological interaction between different layers. For example, the number of inter-layer links is such a structural indicator of coupling. For transport processes, a better measure includes details of how the flows are distributed in the multilayer system and we can define the "interdependency" (Morris and Barthelemy 2012) as

$$\lambda = \frac{1}{T} \sum_{i \neq j} T_{ij} \frac{\sigma_{ij}^{\text{multi}}}{\sigma_{ij}}, \qquad (6.1)$$

where $T = \sum_{ij} T_{ij}$ is the total traffic, and where $\sigma_{ij}^{\text{multi}}$ is the number of shortest paths between nodes i and j that include edges from more than one layer, and σ_{ij} is the total number of paths from i to j. The interdependency belongs to the interval $\lambda \in [0, 1]$ and is dependent on the flow allocation process and not just on the system topology. For example, it can be computed for all possible types of paths, including the minimal or the time-respecting paths (see below). The larger λ, the more one network is relying on the other to ensure efficient shortest path (note that there is usually a maximum value of λ strictly less than one). As we will see below, the interdependency is indeed a crucial parameter that controls the coupling between different transportation modes.

Different types of path

In transportation networks, we can define different types of path. The minimal path is the quickest one and is computed by using the largest speed observed on each link and by neglecting waiting times. These paths represent an unreachable condition, equivalent to having all the existing transportation systems perfectly synchronized for the specific trip under consideration.

Transportation networks are not only weighted networks, but also have an important temporal component. In order to construct a path from A to C through B, the arrival time at B (coming from A) should be less than the departure time from B to C. This type of constraint is of course crucial for transportation networks, but is also fundamental for other processes such as epidemic spread in contact networks (for a review on temporal networks, see Holme and Saramäki 2012). We can thus define a "time-respecting path" as the quickest path which takes into account departure and arrival time constraints given by timetables. In addition for transportation networks, we will also take into account walking times from one mode to the other. The time-respecting path is by definition longer than the minimal path, and as we can see on an example shown in Fig. 6.5, they can be greatly different from each other.

In order to understand the impact of synchronization between different modes, and how far urban transportation systems are from an ideal optimum, we can then compare the quickest time-respecting paths with the minimal path. When doing this comparison between paths, we have to select pairs of origin and

6.1 A multilayer network view of urban navigation

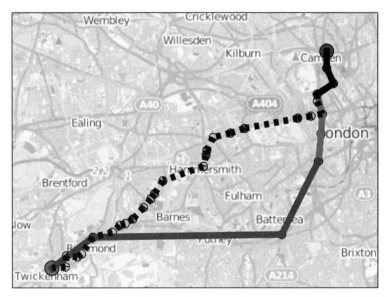

Figure 6.5. Example of the difference between a time-respecting path (solid line) and a minimal path (dashed line). Here, we show a trip from Twickenham to Camden Town in London, and the minimal path would use only buses (continuous line). The optimal time-respecting path in this case is remarkably multimodal: the same bus layer as for the minimal path is used for the final segment, while the rail and metro layers are both used for approaching the city center. Figure taken from Gallotti and Barthelemy (2014).

destination nodes. An important assumption – "the uniform demand" – concerns thus the choice of the origin-destination matrix (see also Chapter 5). Very often, it is unknown and in order to probe the structural properties of these multilayer networks, we assume that origins and destinations are uniformly and independently distributed in the system. In addition, we simplify the problem by not taking into account the departure time. The properties discussed under these assumptions thus constitute an upper bound, obtained by testing all possible pairs of origins and destinations. It is therefore more a structural probe of the system which allows testing of certain features due to intermodal coupling. Obviously all the methods and quantities described here can be very easily adapted to the case where the real OD matrices are known.

Synchronization inefficiency

If the rail and metro layers are relatively quick, they are used for minimal paths, while additional waiting times due to mode change can be too costly for the time-respecting paths. On the other hand, in cities where the bus layer is fast but with a frequency as low as for faster layers (eg. London, Liverpool, Cardiff), minimal paths tend to use buses only, while the time-respecting paths face the synchronization limits of the bus layer itself. More generally, the factors responsible for the time difference between time-respecting and minimal paths are:

- waiting times (both intra- and inter-layer);
- a long walking time for connecting different modes in a given area;
- the fact that optimal riding times used to compute minimal paths may differ from the riding times at a particular hour.

In order to quantify the differences between minimal and time-respecting paths, we introduce the synchronization inefficiency $\delta(i, j)$, computed for two nodes i and j as the ratio of time-respecting travel time $\tau_t(i, j)$ and minimal travel time $\tau_m(i, j)$

$$\delta(i, j) = \frac{\tau_t(i, j)}{\tau_m(i, j)} - 1. \tag{6.2}$$

We can then compute the average over all pairs of nodes

$$\delta = \frac{1}{N(N-1)} \sum_{i \neq j} \delta(i, j) \tag{6.3}$$

where N is the number of nodes considered here for constructing the origin-destination pairs (for example the number of bus stations). The larger this quantity, the less efficient is the multimodal system.

For all cities, δ reaches its maximum δ_{max} for short trips, where waiting times are long compared to the travel time, and then it decreases with the path length ℓ according to the following function, valid for all cities (Fig. 6.6):

$$\delta \approx \delta_{min} + \frac{\delta_{max} - \delta_{min}}{\ell^{\nu}}, \tag{6.4}$$

where $\nu \approx 0.5$. The quantities $\delta_{min}, \delta_{max}$ depend on the city and if we plot $(\delta - \delta_{min})/(\delta_{max} - \delta_{min})$ versus ℓ, we observe a very good collapse for all cities, which suggests that there is an underlying process describing the accumulation of waiting and walking times along time-respecting paths. The specifics of the different cities appear in the system efficiency in both the worst δ_{max} and best δ_{min} limits. We note here that this Eq. (6.4) is consistent with a simple argument based on the central limit theorem leading to $\nu = 1/2$.

Synchronization effect

The differences between the time-respecting and minimal paths find their origin in the lack of synchronization between different modes. It is then natural to quantify this loss in multimodality due to synchronization by the quantity

$$\Delta = \frac{\langle \Lambda_m - \Lambda_t \rangle}{\langle \Lambda_m - 1 \rangle} \tag{6.5}$$

6.1 A multilayer network view of urban navigation

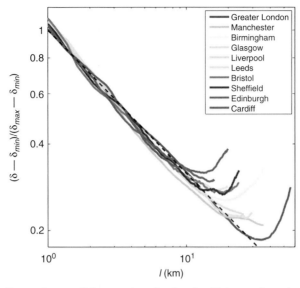

Figure 6.6. Dependence of the synchronization inefficiency δ on the path length ℓ. For all cities, we observe the same function Eq. (6.4), where the specificity of each city is seen in the values of the peak δ_{max} and minimal δ_{min}. The dotted line is a power law fit with exponent ~ 0.5. Figure taken from Gallotti and Barthelemy (2014).

where $\Lambda_{m(t)}$ represents the number of different modes used for minimal trips (m) or time-respecting (t) paths (the brackets denote here the average over all origin-destination pairs). The larger Δ, the larger is the difference between minimal and time-respecting paths. In general, Δ is positive when the average speed of the alternative (ie. non-bus) layers $\overline{V}_{nb}$ is sensibly larger (> 2.5 times) than the average speed $\overline{V}_b$ of the bus layer (Fig. 6.7). When Δ is negative, it usually means that the minimal path is done on the bus layer only and owing to long waiting times, the time-respecting path uses other modes. The quicker the rail and metro layers are, the more multimodal the minimal path would tend to be.

For time-respecting paths, multimodality also implies the importance of synchronization, and it appears that in cities where metro or rail are sensibly faster, their frequency is also lower. In other words, in cities where the fast layers are extremely advantageous in terms of speed, the system suffers from synchronization problems. This empirical finding suggests the existence of a structural limit to the possibilities of transportation systems that policy makers should take into account in the search for an efficient optimization strategy.

Stop events frequency

A crucial question concerns the impact of the number of modes and their frequency on the overall system efficiency. The natural quantity that characterizes this is the

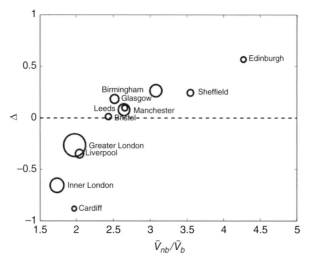

Figure 6.7. The loss in multimodality, due to synchronization in time-respecting path, is related to the speed advantage V_{nb}/V_b from using non-bus layers instead of the bus layer: when the rail or metro layer is fast, it also usually suffers from synchronization problems. Figure taken from Gallotti and Barthelemy (2014).

average number of stop events per unit time

$$\Omega = \frac{\sum_\alpha C_\alpha}{\Delta t} \tag{6.6}$$

where C_α is the number of stop events in the layer α and Δt a given time interval. A good choice consists of taking $\Delta t = 1$ day so that the quantity Ω represents a global measure of the transportation service offered daily in a city. As we will see, this single network indicator Ω is enough to explain the behavior of many key quantities describing the public transport network of different cities.

We first note that Ω (for UK cities as tested in Gallotti and Barthelemy 2014) is increasing with the cities' population (see Fig. 6.8), which seems to indicate that there is indeed an adaptation of the transport offer to the demand.

The interplay between temporal and multilayer aspects of the public transport network is highlighted in Fig. 6.9, showing that the average interdependency $\bar{\lambda}_t$ for time-respecting paths using more than one mode is larger for cities with a larger number of stop events. If we assume that the average number of possible alternatives to the bus-layer path (which is always an available option) is $a_{\bar{\lambda}}\Omega$, the average interdependency of the time-respecting paths is then

$$\bar{\lambda}_t \approx \frac{a_{\bar{\lambda}}\Omega}{1 + a_{\bar{\lambda}}\Omega}. \tag{6.7}$$

6.1 A multilayer network view of urban navigation

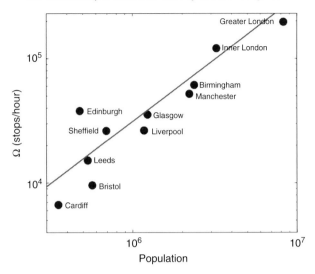

Figure 6.8. The total number of stop events Ω grows proportionally with the urban area population P. The power-law fit $\Omega \propto P^\phi$ gives $\phi \approx 1.0 \pm 0.3$ ($r^2 = 0.88$). Figure taken from Gallotti and Barthelemy (2014).

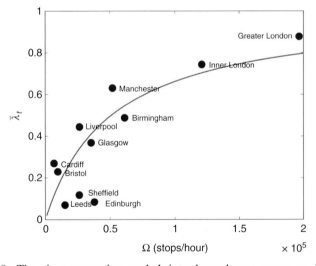

Figure 6.9. The time-respecting paths' interdependency grows as Eq. (6.7), consistent with the hypothesis that the number of possible alternatives to the bus-layer path is proportional to Ω. Figure taken from Gallotti and Barthelemy (2014).

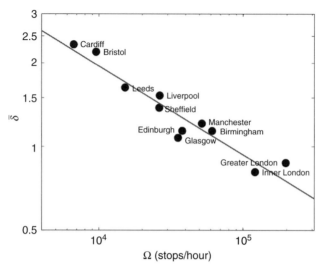

Figure 6.10. The synchronization inefficiency δ decreases with Ω as a power law $\delta \propto \Omega^{-\mu}$, where $\mu \approx 0.3 \pm 0.1$ ($r^2 = 0.91$). Figure taken from Gallotti and Barthelemy (2014).

Using this form to fit the data shown in Fig. 6.9, we obtain $a_{\bar{\lambda}} = 1.65 \times 10^{-5}$ hour per stop ($r^2 = 0.80$).

There are different ways to improve a transportation system such as adding new lines, new connections or increasing the frequencies for a given mode, and they all result in an increase of Ω, which integrates all these modifications, and characterizes the efficiency of a public transportation network in terms of synchronization. Indeed, we observe that the average synchronization inefficiency measure $\bar{\delta}$ decreases with Ω as a power law (see Fig. 6.10)

$$\bar{\delta} \approx \Omega^{-\mu} \tag{6.8}$$

where $\mu \approx 0.3 \pm 0.1$. The expected decrease is naturally due to the fact that larger values of Ω imply larger frequency and thus a better synchronization between modes. The small value of μ is however bad news in terms of efficiency: in order to divide $\bar{\delta}$ by a factor 2 we need to multiply Ω by a factor of almost 10. We can however hope that when exact origin-destination matrices are known, a better optimization of the system can be obtained through targeted improvements.

6.2 The effect of coupling

In this section, we first discuss a simple toy model (Morris and Barthelemy 2012) that allows us to investigate the effect of the coupling on the overall properties of multilayer systems. We then illustrate these theoretical results on a more concrete example of the street network-subway coupling in NYC and in London, showing that in terms of total congestion, there could be an optimal subway velocity for London (Strano et al. 2015).

6.2.1 A toy model

For the sake of simplicity we consider a system such as the street and subway networks. Both these networks are spatial, with nodes and links embedded in two-dimensional space (we will also assume that these networks are planar, which in most cases is a good approximation). In addition, we will consider the case where the layers are connected by a set of nodes common to both networks (*multiplex* case in the terminology defined in Kivelä et al. 2014). In the case of a road network coupled to a subway, all the nodes of the road network are not nodes of the subway network, but conversely, all stations are located at points which can be considered as nodes in the road network.

In order to construct a simple toy model, we generate a first planar network as a triangulation of points in the plane, such as the usual Delaunay triangulation (Guibas and Stolfi 1985), which typically avoids slim triangles that are not seen very often in real networks owing to their inefficiency. We then generate a second spatial network based on a random subset of the points used to construct the first network. This model thus comprises individual networks that are each planar Delaunay triangulations, forming a combined network that is not necessarily planar and where the nodes of the different networks that have the same spatial location are linked together (see Fig. 6.11).

We then need to select the route assignment procedure. A simple choice corresponds to the minimal path, computed as the weighted shortest path. The weight associated with each edge can be chosen to be the length of that edge multiplied by a factor $0 \leq \beta_n \leq 1$, which is common to all edges belonging to the same network. The subscript n is used to label the network: $n = 1$ corresponds to the larger network and $n = 2$ to the smaller. The idea is that $\beta = \beta_2/\beta_1$ is a single parameter that controls the relative weight per unit distance between the two networks. Indeed, in order to simplify further, we impose the artificial constraint that $\beta \leq 1$. This has the effect that a journey on the smaller network ($n = 2$) is favored over a journey of equivalent distance on the larger network ($n = 1$). More concretely, β can be considered as the ratio of average velocities in the different layers.

The number β controls the coupling between the two networks, and we can consider how well the system is operating when varying β. For example, we can

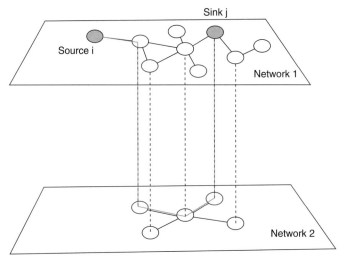

Figure 6.11. A system made of two coupled networks where the nodes of network 2 form a subset of the nodes of network 1. Nodes common to both networks are considered to be coupled to each other (the coupling is shown by dashed lines). We represent a path (light gray line) using the two layers between two nodes, the "source" or "origin" i, and the "sink" or "destination" j. Figure taken from Morris and Barthelemy (2012).

test its effect on the average time traveled

$$\bar{d} = \frac{1}{T} \sum_{i \neq j} T_{ij} d_{ij}, \qquad (6.9)$$

where d_{ij} is the weighted distance (or "time") traveled between nodes i and j on the network and T_{ij} is the OD matrix. It is reasonable to think that a well-designed system reduces the average time traveled (i.e., water/food supply, the Internet, transportation, etc.).

Another important quantity, which is a simple proxy for traffic, is the edge betweenness centrality, defined as (see section 4.2)

$$g(e) = \sum_{i \neq j} T_{ij} \frac{\sigma_{ij}(e)}{\sigma_{ij}} \qquad (6.10)$$

where subscript e is used to label edges, σ_{ij} is the number of shortest paths between i and j, and $\sigma_{ij}(e)$ is the number of shortest paths between nodes i and j, which use edge e. The betweenness centrality allows us to introduce another measure based on the Gini coefficient G, which is a number between zero and one. This quantity G is typically used in economics for the purpose of describing the distribution of wealth within a nation, and here we use it to characterize the

6.2 The effect of coupling

disparity in the assignment of flows to the edges of a network, something that has been done before for transportation systems such as the air traffic network (Reynolds-Feighan 2001). For example, if all flows were concentrated onto one edge, G would be one, whilst if the flows were spread evenly across all edges, G would be zero. We use the definition according to (Dixon et al. 1987)

$$G \equiv \frac{1}{2E^2 \bar{g}} \sum_p \sum_q |g(p) - g(q)|, \qquad (6.11)$$

where subscripts p and q label edges, E is the total number of edges, $g(p)$ is the flow assigned to edge p as defined earlier, and $\bar{g} = \sum_p g_p / E$ is the average flow on an edge.

The last thing that remains is to allocate flows on the network, defined by an origin-destination (OD) matrix T_{ij}. It is clearly impossible to consider the whole space of all OD matrices and we choose a specific example, interpolating from a monocentric matrix to a random graph, which allows understanding of the impact of flow structure. More precisely, for each node with a probability p, we connect it to a random destination, and with a probability $1 - p$, it connects to the origin (see Fig. 6.12).

We consider this as a simple model of two transport modes coupled to each other (and on the road network we assume that the congestion does not affect route choice) and we vary the ratio $\beta = \beta_1 / \beta_2$ of the two inverse velocities of the layers 1 and 2, and observe the overall behavior of the system. In this picture, the

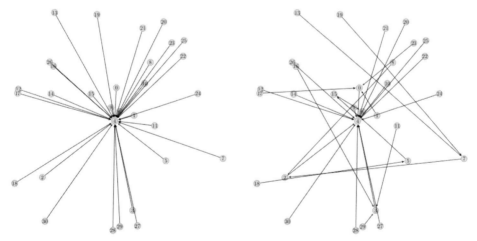

Figure 6.12. Representations of OD matrices where each arrow corresponds to an entry in T_{ij}. (Left) Monocentric case. (Right) A non-monocentric case obtained by rewiring links of the monocentric OD with probability $p = 0.5$. Figures taken from Morris and Barthelemy (2012).

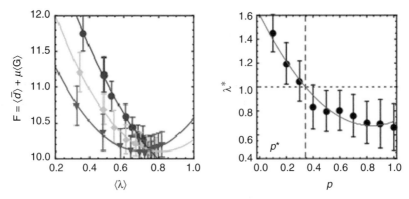

Figure 6.13. Existence of an optimal coupling: (Left) Simulation results for $\mu = 10$, $N_1 = 100$, $N_2 = 20$, and p values: 0 (dots), 0.4 (diamonds), and 0.8 (inverted triangles) (three values only of p are shown to ensure the lines of best-fit can be seen clearly). (Right) Minima of quadratic best-fit curves for different values of p. We obtain $\lambda^* = 1$ for $p^* \simeq 0.34$. Figures taken from Morris and Barthelemy (2012).

Gini coefficient can be thought of as a measure of road use. A low value indicates that the system uses all roads to a similar extent, whilst a high value indicates that only a handful of roads carry all the traffic. With this analogy in mind, it is natural to combine the effects observed above into a single measure. We assert that it is likely a designer or administrator of a real system would wish to simultaneously reduce the average travel time and minimize the disparity in road utilization. To serve this purpose, we define a "utility" function

$$F = \langle \bar{d} \rangle + \mu \langle G \rangle \tag{6.12}$$

which measures the total efficiency of the system (μ is a positive constant). Numerical simulations of this system lead to the result shown in Fig. 6.13(left) where we observe that a non-trivial (ie. non-maximal) optimum λ^* emerges. Whether a non-trivial optimum coupling exists depends on the origin-destination matrix (see Fig. 6.13(right)). For OD matrices rewired with a high probability, increasing the speed of the rail network reduces the road utilization as flows become concentrated around nodes where it is possible to change modes. Dependent on the value of μ, the effect of reduced utilization can outweigh the increased journey time, leading to a minimum in F. Monocentric OD matrices, by contrast, have inherently inefficient road utilization when applied to planar triangulations, regardless of the speed of the rail network. Therefore no minimum is observed, and hence no (non-trivial) optimum λ^*. More systematically, we plot the minima λ^* of best-fit curves corresponding to different values of p (Fig. 6.13(right)). Defining p^*, the value of p for which $\lambda^* = 1$, it is possible to categorize the system into one of two regimes. We observe that:

- if $p < p^*$, then the optimal coupling is trivially the maximum;
- otherwise if $p \geq p^*$, a non-trivial optimal coupling exists, and hence a non-trivial value of the velocity ratio β.

This simple toy model is thus characterized by two competing "forces," the desire to move all flows onto the most efficient network, whilst also ensuring that congestion does not arise around the nodes that connect both networks. We observe that the optimization of such a system can be sensitive to randomness introduced in the origin-destination matrix and thus to the demand.

6.2.2 Optimal velocity for the road–subway system

In the previous section we considered a toy model, highlighting the effect of the coupling between two different transportation modes. We present here the same analysis, but on real street and subway networks for the two large metropolitan areas of London and New York (Strano et al. 2015), which are very different in size and geography as can be observed in Fig. 6.14(a,b). In particular, we analyze the effect of varying the subway speed and show that increasing it can lead to unexpected counter-effects.

The data (obtained from www.openstreetmap.org) give the weighted graph $G_s = (V_s, E_s)$ of the connected street network in its "primal" representation, with nodes being street junctions and edges representing the street segments connecting them.

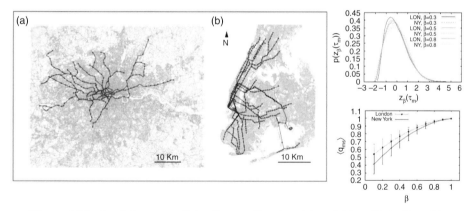

Figure 6.14. Spatial extent of the Greater London (a) and New York City (b). Note that the Greater London area is not covered by the underground system, in contrast to New York where most areas are connected by the subway. (c) Distribution of normalized quickest path times z_β (see Eq. (6.13)). (d) The quantity $\langle q_{ms} \rangle$ averaged over all nodes as a function of β (the error bars indicate here the dispersion around the average). The average ratio of travel times with and without the subway layer is typically of order 0.5 and does not vary much with β. Figures from Strano et al. (2015).

Similarly, we obtain the connected subway network $G_u = (V_u, V_u)$ with nodes representing underground stations and links connecting successive stations on the same line. The multilayer network G_{multi} is then naturally defined as the union of these two networks.

For both networks, the weights of the links are given by their length and subway stations, and road intersections are here considered as being different nodes. Subway stations are accessible from more that one access on the street, but for the sake of simplicity, the multilayer network is constructed (Strano et al. 2015) by connecting each subway station to its closest street junction only (a simplification that won't change the structure of quickest paths).

The generic nature of quickest paths

New York is composed of two large and almost disconnected components with the subway system covering a similar spatial extent and carving-up the different boroughs. London instead presents – at a large scale – a typical radiocentric urban structure with the underground systems connecting satellite districts and peripheries to the urban core. Differences both in size and geography between these cities are also reflected by basic network descriptors such as the number of nodes, etc. (Strano et al. 2015).

We assume that the efficiency of the subway is due to its speed, which is in general higher than that of overground modes such as private cars, taxis, or buses. In order to reflect this, we introduce a parameter $0 < \beta \leq 1$ that describes the ratio of speeds in both systems, similarly to the toy model proposed in the previous section. A smaller β corresponds to a faster subway speed, as compared with the speed on the street network. The introduction of this parameter allows us to study the properties of the multilayer system as a function of subway speed. We note that this parameter can be measured empirically: for London we have $\beta_{London} \approx 0.48$, while for NYC we observe a slightly larger value, $\beta_{NYC} \approx 0.55$.

We denote by $\tau_s(i, j)$ the travel cost (i.e., the number of time units) of the quickest path between street nodes $i, j \in V_s$, and by $\tau_m(i, j)$ the cost of the quickest path between i and j in the multilayer network (which can traverse *both* street *and* subway links). The normalized quantity

$$z_\beta(\tau_m) = \frac{\tau_m - \langle \tau_m \rangle}{\sqrt{Var(\tau_m)}}, \quad (6.13)$$

displays a behavior that is roughly constant for β larger than 0.2–0.3, as shown in Fig. 6.14c, demonstrating that the effect of β is essentially contained in the average and variance of τ_m. This is a rather surprising result, given that the two cities display many geographical and structural differences. The cost $\tau_m(i, j)$ between

6.2 The effect of coupling

nodes i and j can be written as

$$\tau_m(i, j) = \sum_{e \in P(i,j)} \tau(e) \quad (6.14)$$

where the sum is over all links e that belong to the quickest path $P(i, j)$ and where $\tau(e)$ is the cost on this link (we neglect inter-modal change costs in this simple argument). If the path is long enough, if the random variables $\tau(e)$ do not display long-range correlations and are not broadly distributed, the central limit theorem applies and the distribution of τ_m follows a Gaussian distribution in a certain range and therefore provides an explanation for the seemingly "universal" behavior of z_β. There are obviously deviations observed for small values of β arising from the fact that the durations of the paths become very heterogeneous depending on the proximity of their origin or destination to subway stations. In this respect, a very high relative subway velocity enhances spatial differences in the city and may lead to an uneven distribution of accessibility, a fact that will be confirmed below with the local outreach analysis.

Another interesting quantity is the average ratio between the travel costs from i to other street nodes through the multilayer network and through the street network

$$q_{ms}(i) = \frac{1}{N_s - 1} \sum_{j \in V_s} \frac{\tau_m(i, j)}{\tau_s(i, j)} \quad (6.15)$$

where N_s is the number of street nodes. The smaller this ratio, the larger the effect of the subway on travel costs. We see in Fig. 6.14d that typical values are of order 0.5 for both cities and that the effect of β is rather weak: a decrease from $\beta = 1$ to $\beta = 0.5$ leads to a decrease in $\langle q_{ms} \rangle$ of order 20%. In addition, it seems that the effect of subways in London is less important than in New York, which is probably due to the lesser extent of the underground in the Greater London area.

A central quantity for describing the importance of inter-modality is given by the interdependency for each pair of nodes

$$\lambda(i, j) = \frac{\sigma_{ij}^{\text{multi}}}{\sigma_{ij}} \quad (6.16)$$

where σ_{ij} is the total number of shortest paths between i and j (using either one or two networks), and $\sigma_{ij}^{\text{multi}}$ the number of paths using edges of both networks at least once. If we sum over all possible destination nodes j, we can quantify the added value of the interlayer coupling to the reachability of nodes, and obtain the interdependency of a street node $i \in V_s$ defined as

$$\lambda(i) = \frac{1}{N_s - 1} \sum_{j \in V_s} \frac{\sigma_{ij}^{\text{multi}}}{\sigma_{ij}}. \quad (6.17)$$

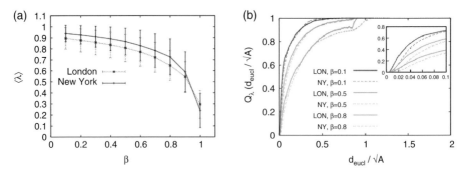

Figure 6.15. (a) Average interdependence, $\langle\lambda\rangle$ as a function of β. (b) Normalized interdependence profile computed for different values of β. Both cities exhibit a similar behavior despite very different geographical structures. Figure from Strano et al. (2015).

When averaged over all nodes, we then obtain $\langle\lambda\rangle$ defined in the previous sections.

In order to understand the effect of scale on the interdependence, we also define the interdependence profile as

$$Q_\lambda(d) = \frac{1}{N(d)} \sum_{\substack{i,j \in V_s \\ d_e(i,j)=d}} \lambda(i,j) \qquad (6.18)$$

where $d_e(i, j)$ is the Euclidean distance between i and j and $N(d)$ is the number of pairs of nodes at Euclidean distance d. In Fig. 6.15a we show the average interdependence among all street nodes as a function of β and the resulting interdependence profile Fig. 6.15b. We see from these figures that, in both cities, the existence of the subway has a very large impact. For example, for $\beta = 0.8$ we obtain $\langle\lambda\rangle$ around 0.7, meaning that even when the subway is only 1.25 times faster than the street network, already about 70 percent of the quickest paths are going through the subway. A slight decrease of β from 1 has thus a large impact on the structure of quickest paths, while for smaller values of β, improving the subway speed does not bring a significant improvement in the structure of quickest paths. In addition, as shown in Fig. 6.15b, we observe in both cities a sharp increase in $Q_\lambda(d)$ for small Euclidiean distances, meaning that already for relatively short trips, it is worth "hopping on" the subway (note that we neglect here waiting, walking, and connecting times which can be significant, as we saw in the previous section). The slope of the interdependence profile (see inset of 6.15b) at small $d_{eucl} \simeq 0$ is increasing as β is decreasing, suggesting that a slight increase in subway speed could make the networks highly interdependent even at very small scales.

6.2 The effect of coupling

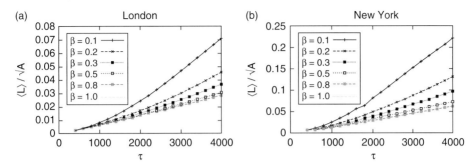

Figure 6.16. Average local outreach $\langle L \rangle$ normalized by the square root of area, for London (a) and New York (b). Figures taken from Strano et al. (2015).

Outreach and urban horizon

The presence of a transportation mode such as a subway affects the overall performance of a city in terms of efficiency of transport and the accessibility of certain locations, but also has an important impact on how pairs of locations are connected. In order to measure this effect, we define the *spatial outreach* of a street node $i \in V_{\text{street}}$ as the average Euclidean distance from i to all other street nodes that are reachable within a given travel cost τ

$$L_\tau(i) = \frac{1}{N(\tau)} \sum_{j | \tau_m(i,j) < \tau} d_e(i,j) \qquad (6.19)$$

where $N(\tau)$ is the number of nodes reachable on the multilayer network within a given travel cost τ. In Fig. 6.16 we show the average local outreach as a function of the travel cost threshold τ, which displays a non-linear behavior due to the different speeds achievable in the two transportation modes. This provides support for a general effect already known: for longer trips, faster transportation modes are used (see Section 5.3 and Gallotti and Barthelemy (2014) for the UK case). The average velocity depends obviously on β, but apparently not in a simple linear way.

As shown is Fig. 6.17(a-d), as the subway velocity increases, the nodes having a high local outreach are concentrated close to subway stations where the subway is the most accessible, and the graph consisting of high-outreach nodes becomes less fragmented. In other words, as the underground becomes faster, a continuous area of high-outreach nodes emerges in the city center and around the nodes of the subway network, implying that an individual can travel from this area to far-away places at a small travel cost τ. The location of this highly accessible zone cluster from a dispersed configuration (Fig. 6.17d) to a centralized one (Fig. 6.17a), which shows a centralization effect due to the accessibility provided by the subway.

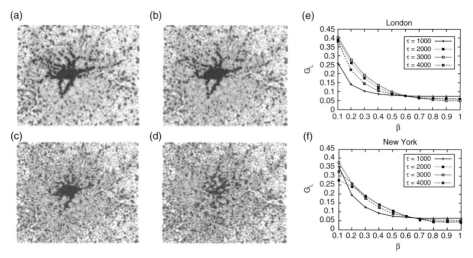

Figure 6.17. (a-d) Maps showing for London the spatial distribution of the local outreach for a travel cost $\tau = 2000$, and speed ratio (for a-d: $\beta = 0.1$, $\beta = 0.7$, $\beta = 0.4$, $\beta = 1$). The darker the node, the larger is its outreach. As the Underground's speed increases compared to the speed of the street network, the nodes having a high local outreach are concentrated at the center along the Underground network, where the Underground is most accessible. (e, f) The graphs show the Gini coefficient of the local outreach versus β for both cities. Figure from Strano et al. (2015).

In order to characterize the heterogeneity of the local outreach L, we compute the Gini coefficient G_L from its distribution. The results are shown in Fig. 6.17(e,f) and we see that for both cities for $\beta > 0.5$ the accessibility is distributed almost uniformly amongst all the places in the cities, while for a faster Underground the shift to an uneven distribution of accessibility is clear. This result suggests that transportation policies that focus on increasing the speed on a single travel modality may lead to undesirable spatial heterogeneity in the accessibility of different locations.

For given values of β and τ there is a maximal value L_m which corresponds to the largest outreach for a given trip duration and subway efficiency. We can estimate this value by using the following simple argument: the maximum value is reached when the path is made on the quickest transportation mode, which is the subway (with velocity v/β where v is the average car velocity). The probability of having a station within walking distance d_0 (see also the coverage discussed in Section 4.4) is then

$$p = \rho_u \pi d_0^2 \qquad (6.20)$$

where $\rho_u = N_u/A$ is the density of subway stations ($A = L^2$ is the area of the city and N_u is the number of subway stations). The maximal outreach L_m is then

given by
$$L_m = \frac{N_u}{L^2}\pi d_0^2 \frac{v}{\beta}\tau. \tag{6.21}$$

The maximal fraction of the city reachable in the time τ is $\alpha_c = L_m/L$ and is then given by
$$\alpha_c = \frac{N_u}{L^3}\pi d_0^2 v \frac{\tau}{\beta}. \tag{6.22}$$

This last equation shows in particular that the quantity $\alpha_c L^3/N_u$ should increase linearly with τ/β, with a constant of proportionality depending on the geometry of the city. As shown by Strano et al. (2015), this scaling is in good agreement with simulation, showing that this simple argument captures some important aspects of this phenomenon.

The geography and distribution of urban centrality

As discussed in Section 4.2, the spatial distribution of the BC contains a lot of information about the network and its spatial properties. For multilayer networks we can think of many different measures (see for example De Domenico et al. 2013; Kivelä et al. 2014; Boccaletti et al. 2014) and we examine here the somehow simpler question of how the spatial distribution of the BC in the street network is affected by the subway system. The BC of a street node $v \in V_s$ in the street network is defined as
$$bc_s(v) = \frac{1}{(N_s-1)(N_s-2)}\sum_{i\neq v\neq j \in V_s}\frac{\sigma_{ij}^{\text{street}}(v)}{\sigma_{ij}^{\text{street}}} \tag{6.23}$$

where $\sigma_{ij}^{\text{street}}$ is the number of quickest paths between i and j in the street network, of which $\sigma_{ij}^{\text{street}}(v)$ goes through street node v. Similarly, the betweenness centrality of a street node $v \in V_s$ in the multilayer network is
$$bc_m(v) = \frac{1}{(N_s-1)(N_s-2)}\sum_{i\neq v\neq j \in V_s}\frac{\sigma_{ij}^{\text{multi}}(v)}{\sigma_{ij}^{\text{multi}}} \tag{6.24}$$

where $\sigma_{i,j}^{\text{multi}}$ is the number of quickest paths between i and j in the multilayer network, of which $\sigma_{ij}^{\text{multi}}(v)$ go through street node v.

The maps in Fig. 6.18(a,b,c,d) show the BC spatial distribution for both cities computed on streets for two different values of $\beta = 1$ (a,b) and $\beta = 0.1$ (b,c).

These maps clearly display a dramatic change in the spatial distribution of central places when introducing a subway system, shifting congestion from internal street routes and bridges to inter-modal places located at the terminal points of the subway networks, which presumably are used as entry/exit gates for suburban flows to reach core urban areas. Remarkably enough, in both cities

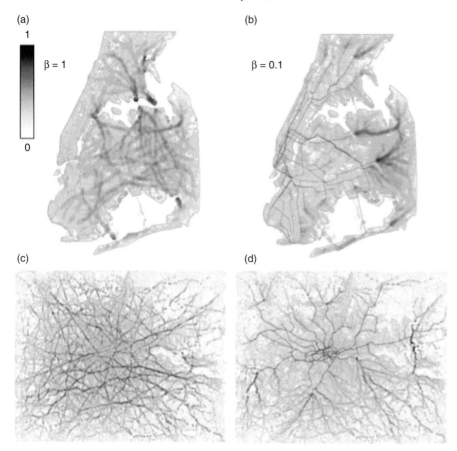

Figure 6.18. The spatial distribution of BC on the New York (top) and London (bottom) street network for two different values of β. We observe a clear crossover from congested road locations for $\beta = 1$ (left column) to "focal points" of the subway system for small β (right column). Figures taken from Strano et al. (2015).

these places are located in urban areas that do not overlap with the subway system, thus possibly creating congestion in unexpected places. In other words, the introduction of subway networks operates as a decentralizing force, creating congestion in places located at the ends of underground lines and not, for example, in the city center as one might expect referring to classical results on rewiring processes for chain or lattice networks, in which the BC is correlated to the distance to the barycenter of nodes (Barthelemy 2011). The statistical dispersion of BC can be measured by its Gini coefficient and also suggests that congested places always become more critical in the system as β decreases. In fact, as shown in Fig. 6.19, the Gini coefficient for the BC increases as the subway becomes more efficient (faster, decreasing β), meaning that a larger fraction of quickest paths use it. The BC distribution is then less homogeneous, making the system more

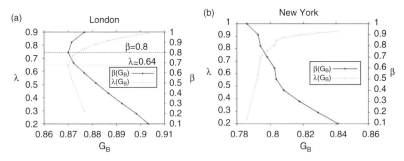

Figure 6.19. β and interdependency λ as a function of the dispersion of the multilayer BC measured by the Gini coefficient G_B for the BC.

fragmented and less resilient. Examining the BC Gini as a function of β and the interdependency λ in London (Fig. 6.19a), we observe a non-trivial optimal value for β for which flows are the most homogeneously distributed across street junctions. In New York (Fig. 6.19b), however, there seems to be no optimum, which suggests (according to the theoretical discussion of the previous section) that – surprisingly – it has a more marked monocentric aspect than London. In other words, the congestion in central places in New York is so large that introducing an efficient subway system is always better, even if it creates congestion at other points. These results on the BC and on the existence of an optimal point are thus in agreement with the results of the simple toy model of coupled transportation networks (Section 6.2.1), where – depending on the distribution of trip targets – two regimes were observed: one in which the optimal coupling is trivially the maximum, and another one where a non-trivial optimal coupling exists.

Obviously, the results presented here certainly depend on the structure of flows and the choice of OD matrices. A more precise study should then involve the real OD matrices, but the results here highlight the facts that in coupled networks, improvements on one layer are not necessarily globally positive, and that the whole system should be considered in this type of transportation analysis.

6.3 Information perspective on navigation in cities

As we saw in the previous sections, the growth of large urban areas usually goes together with the development of transportation infrastructure and an increase in the number of modes. The growth of such large areas leads to transportation systems that are so entangled and complex, that it is natural to ask how to measure this complexity and if it is above human cognitive limits (Gallotti et al. 2016b). In this section, we discuss a way to measure this complexity, based on the information entropy. Using this measure on the world's largest subway systems, we identify the cognitive limit, which is of order 8 bits and which corresponds to the most complicated trip that can be stored in our visual working memory. Using this limit, we then estimate the complexity of multilayer networks for three megacities (New York, Paris, and Tokyo) and show that most of the multimodal trips exceed this limit. Multimodal transportation networks became so complex that a visual representation on a map is actually too complex and useless. This implies that our traditional view of navigation in cities has to be substantially revised.

6.3.1 Simplest paths

When we ask for directions in a city, the answer is often in the form of a sequence of roads that we will have to take. It is thus natural that we will not discuss the properties of the road network in primal space but in the dual, where the nodes represent straight lines (or roads) and the links represent intersections between roads. The shortest path in this dual space then corresponds to the path with the smallest number of turns (and if there is more than one such path, we choose the shortest one) and is called the "simplest path." For example, we show the regular lattice in Fig. 6.20 in both the primal and dual representations. The dual network has a diameter equal to two, meaning that at most two turns are necessary to connect any pair of routes.

6.3.2 Information entropy

In order to characterize the complexity of paths we use the entropy, which is a measure of missing information. Generally speaking, if we have a probability p_i

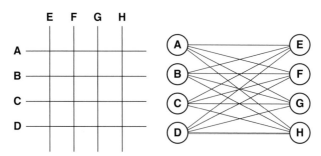

Figure 6.20. Primal (left) and dual (right) networks for a square lattice. In this example, the lattice has $N = 8$ routes. Each route has $k = N/2 = 4$ connections, so the total number of connections is $K_{tot} = k^2 = 16$. In the dual network, the four East–West routes (A,B,C,D) and the four North–South routes (E,F,G,H) yield a graph with a diameter of 2. Figure taken from Gallotti et al. (2016b).

6.3 Information perspective on navigation in cities

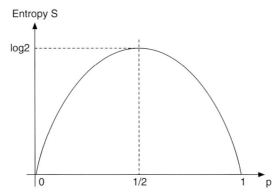

Figure 6.21. Entropy for the simple bimodal case. For $p = 0$ and $p = 1$ the entropy is zero, corresponding to the minimum of missing information. The maximum entropy at $p = 1/2$ corresponds to the maximal uncertainty and to the maximal information gained from the observation of the outcome of the random process.

that a system is in state i, the entropy is defined as

$$S = k \sum_i p_i \log p_i, \quad (6.25)$$

where the constant k sets the scale and units. If we choose $k = 1$ and the logarithm in base 2, the entropy is in *bits*. This quantity measures the missing information that corresponds to the distribution of the p_i's. In other words, if the system is governed by this distribution, when we observe an actual realization of the system, we will increase our knowledge by a certain amount of information, and this is what the entropy measures. For example if the distribution is not random, only one of the p_is is non-zero, and the entropy is zero: we gain no information when observing the system. For a more complex case such as a bimodal distribution we obtain the curve shown in Fig. 6.21. In this figure, we see that when $p = 0$ or $p = 1$, the system is in one given state and no information is gained when observing an actual realization of the system. The maximum entropy is obtained for $p = 1/2$, which corresponds to the case where both states are equiprobable and the observation of the system brings us maximal information.

Rosvall et al. (2005) proposed to apply entropy to the complexity of paths in street networks. More precisely, they measure the information that is needed to encode a shortest path (in dual space) from a route s to another route t. The amount of necessary information can depend strongly on the initial and final nodes, and we consider here a trip from an origin node i in route s to a destination node j in route t. This trip is embedded in real space, and among all possible simplest paths (Rosvall et al. 2005) we pick the fastest one $p(i, s; j, t)$. The total information for knowing such a path is

$$S(i,s;j,t) = -\log_2\left(\frac{1}{k_s}\prod_{n\in p(i,s;j,t)}\frac{1}{k_n-1}\right), \qquad (6.26)$$

where $p(i,s;j,t)$ is the sequence of routes needed for connecting i in route s to j in route t. The term k_s is the number of routes connected to s, and along the path we have the choice between $k_n - 1$ routes. The idea behind Eq. (6.26) is that when tracking a trip along a map (with the eyes or a finger), the connections that one has to exclude represent – similarly to the number of distractors in visual search tasks (Credidio et al. 2012) – the information that has to be processed and thus temporarily stored into working memory (Baddeley 2003). One can therefore construct the measure of entropy (6.26) as a proxy for the accumulated cognitive load that is associated with the trip, and is analogous to the total amount of load experienced during a task (Paas et al. 2003).

From a map user's perspective, the existence of several alternative simplest paths is not necessarily a significant factor, as one only needs a single simplest path for successful transportation from origin to destination. Consequently, we use the entropy in Eq. (6.26) rather than the one proposed by Rosvall et al. (2005). To produce a single summary statistic for a path, we average $S(i,s;j,t)$ over all nodes $i \in s$ and $j \in t$ (denoted by brackets $\langle \cdot \rangle$)

$$\bar{S}(s,t) = \langle S(i,s;j,t)\rangle. \qquad (6.27)$$

This is the main quantity that we will use in the following in order to describe the complexity of a trip and which allows extraction of an empirical upper limit to the information that a human is able to process for navigating in a map.

6.3.3 Information threshold: 8 bits

The values of $\bar{S}(s,t)$ in a network tend to grow with the number C of connections that appear in a simplest path and with the average degree $\langle k \rangle$ of the nodes in dual space as well. Adding new routes can thus have a negative impact from the information perspective: although new routes can be useful for shortening the simplest paths for some (s,t) pairs, new connections simultaneously increase the average degree of a network and can make it more difficult to navigate in it. We thus want to estimate the maximum possible information that an individual can reasonably process in order to navigate in a transportation system. For that purpose, we consider the world's 15 metro networks with the largest number of stations (see for example Section 4.3 for some characteristics of large metro networks). For each network, we consider the shortest simplest paths with $C = 2$ connections. This corresponds to paths that use 3 different lines: such a path starts

6.3 Information perspective on navigation in cities

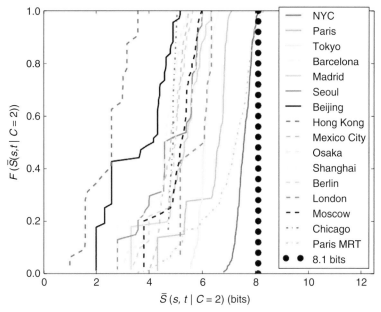

Figure 6.22. Cumulative distribution of the information needed to encode trips with two connections in the 15 largest metro networks. The largest value occurs for the New York City metro system, which has trips with a maximum of $S_{max} \approx 8.1$ bits (represented by the vertical dotted line). Among the 15 networks, the Hong Kong and Beijing metro networks have the smallest number of total connections and need the smallest amount of information for navigation. The Paris MRT (Metro, Light Rail, and Tramway) network from the official metro map (www.ratp.fr) includes three transportation modes (which are managed by two different companies) and reaches values that are similar to those in the larger NYC Metro. Figure from Gallotti et al. (2016b).

from a source route s, connects to an intermediate route r, and then connects to a destination route t. In primal space, this corresponds to four points: the origin, the destination, and two connecting points. We have two motivations for this choice:

- it reflects the limited ability of humans to keep information on a maximum of about four objects (origin, destination, and two connection points) in visual working memory (Luck and Vogel 1997);
- in most of the 15 cities, two connections correspond to the diameter of the dual network and therefore represent the most complex trip that can occur on these systems.

In Fig. 6.22, we show the cumulative distribution of entropies $\bar{S}(s,t|C=2)$ for these 2-connections paths. We find that the New York City metro system is the largest and most complex metropolitan system in the world, with a maximal value of $S_{max} \approx 8.1 \approx \log_2(274)$ bits. Paris' transportation system reaches a similar value if one takes into account the light rail and tram system as in the official metro map. Navigation in such large networks is already non-trivial, and it has been observed (see Blascheck et al. 2014 and references therein) that there is an eye-movement behavioral transition when the system becomes too large (i.e., when there are too many connections). The value S_{max} for trips with two connections thus provides a natural limit, above which human cognitive capabilities are challenged and for which it becomes extremely difficult to find a simplest path. We thus make the reasonable choice to take S_{max} as the cognitive limit for public transportation: a human needs information entropy of $\bar{S}(s,t) \leq S_{max}$ to be able to navigate in a network successfully without assistance from information technology tools. We note here that this limit does not take into account other factors such as training or spatial information. Knowing a city and its metro map well, using spatial information to discard some lines, will reduce the difficulty of the route-searching task. We consider here the upper limit, where an individual discovers a network for the first time and has to find her way in it.

To gain a physical understanding for the cognitive limit S_{max}, we estimate $S(s,t|C=2)$ for a regular lattice (like the one in Fig. 6.20) with N lines that are connected with $N/2$ other lines (i.e., $k_r = N/2$ for all r). This choice of lattice is justified by the results of Roth et al. (2012), showing that most large metropolitan transportation networks consist of a core set of nodes with branches that radiate from it (see also Section 4.3). The core is rather dense and has a peaked degree distribution, and it is thus reasonable to use a regular lattice for comparison. In the dual space of the regular lattice, the degree k_s of route s is equal to $N/2$, and we thus obtain

$$\bar{S}(s,t|C=2) = \log_2[k_s(k_r-1)]$$

$$\approx \log_2(\langle k \rangle^2) = \log_2\left(\sum_{i=1}^{N} k_i/2\right), \quad (6.28)$$

where $\langle k \rangle$ denotes the average degree. The last equality in Eq. (6.28) comes from the relation for the total degree of a regular lattice: $\sum_{i=1}^{N} k_i = \langle k \rangle N = 2\langle k \rangle^2$. The key quantity for understanding S_{max} is therefore the total number of undirected connections $K_{tot} = \sum_{i=1}^{N} k_i/2$ in dual space. As we indicated in Eq. (6.28), this is identical to the square of the average degree $\langle k \rangle^2$ in a lattice. For Paris, for example, we obtain $\langle k \rangle \approx 9.75$, which leads to $9.75^2 \approx 95$ connections for the corresponding lattice. The actual Paris metropolitan network has a total of 78 connections, and the difference comes from the fact that the real network is not a perfectly regular lattice.

6.3.4 Effect of multimodal couplings

Once we have estimated the cognitive threshold for the most complex paths in the 15 largest metropolitan networks, we can now consider the effect of including other

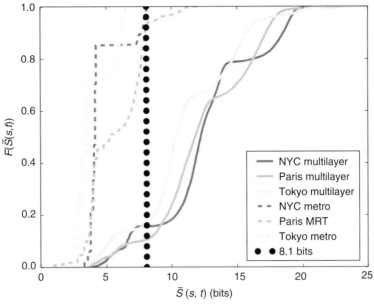

Figure 6.23. Information entropy of multilayer networks. The solid curves represent the cumulative distributions of $\bar{S}(s,t)$ for multilayer networks that include a metro layer for New York City, Paris, and Tokyo. (We associate one layer with bus routes and another with metro lines.) Most of the trips require more information than the cognitive limit $S_{\max} \approx 8.1$. The fraction of trips under this threshold are 15.6% for NYC, 10.7% for Paris, and 16.6% for Tokyo. The dashed curves are associated with all possible paths in a metro layer; in this case, the amount of information is always below the threshold, except for Paris, which includes trips with $C = 3$. Note that the threshold value lies in a relatively stable part of all three cumulative distributions, which suggests that our results are robust with respect to small variations of the threshold choice. Figure taken from Gallotti et al. (2016b).

transportation modes. In order to do this, we compute the information entropy for simplest paths estimated over all the multiple transportation layers, for the three megacities New York City, Paris, and Tokyo. We then see (see Fig. 6.23) that essentially more than 80% of the trips in the multimodal transportation network of these major cities require more information than the most complicated trajectory in the largest metro networks (the other 20% correspond to pairs of nodes for which the trip has essentially one connection or two connections but starting from a small degree route).

The number of connections acting as distractor for the case of the Paris MRT is already so large that it has a crucial impact on the route search (it takes on average ≈ 30 seconds for such a search; see Blascheck et al. 2014), and the complexity of the bus layer (and therefore of the coupled metro and bus system) will therefore exceed human capacity. Consequently, traditional maps that represent all existing bus routes have very limited utility. This result thus calls for the need to think about a user-friendly way to present and to use bus routes. For example, unwiring some bus–bus connections reduces the information, and leads to the idea that a design centered around the metro layer could be efficient. Further work is however needed to reach an efficient, "optimal" design from a user perspective.

We have to stress the fact that this calculation corresponds here to the worst case, where we do not have extra information about the path. It is clear that in general we do have such extra information as the spatial location of the destination. This allows us to discard some routes and to reduce the degree at some points. The final result of having more information is then to reduce the effective entropy of paths. It would be interesting to integrate this type of effect and see how it could improve our navigation possibility in maps.

More generally, this result suggests the unsettling perspective that very soon it will be very difficult to navigate in a large city with only a map and that ICT tools will become necessary. Even if we do have such tools, it is however important to realize that the complexity of our cities might exceed the boundaries of what is within our intuitive capacities. This probably will push us to rethink how we understand and visualize a city and its complexity.

7
Socioeconomic aspects

Understanding the main social and economical indicators of cities represents a formidable task and we can essentially hope at this stage to understand a certain number of various aspects. It indeed seems out of reach to propose a general theory that could integrate all these aspects at once, but rather we will focus on specific ones, without forgetting of course that in a city many features are connected to each other. We will discuss simplified approaches and again, we cannot pretend that these models are the end of the story, but the goal in each case is to identify the relevant parameters and the dominant mechanisms.

Urban economics produced a number of models that now constitute the current basis for the theoretical understanding of cities (see for example the 1989 textbook by Fujita). The spatial structure of rent, of income, the impact of transportation systems or amenities, are all subjects discussed in a theoretical framework developed in urban economics. For this reason, it is important to know these models and their main ingredients identified by economists, as they can serve as a basis for other approaches.

In this chapter, we start by discussing the important Alonso–Muth–Mills (AMM) model and its extension to different transportation modes. We will also discuss the Beckmann model as it proposes to integrate social interactions, an ingredient for which we have always more data. For these cases we will provide a complete, self-contained discussion which could serve as an entry point to this field of research for non-economists.

We will then discuss the important problem of the spatial organization of income, in particular how to quantity the segregation level in empirical data. From a theoretical point of view, this problem was discussed in the famous Schelling model and we will consider here physical approaches to this model.

Finally, we will discuss empirical work about scaling in cities of important indicators. Measures on macroscopic quantities revealed (Pumain 2004; Bettencourt et al. 2007) a scaling behavior of the form $Y \sim P^\beta$, where the exponent β

can be different from 1. These results triggered the interest of many scientists as non-trivial values of β might provide a guide for testing models, akin to critical exponents for statistical physics (see for example Stanley 1971).

7.1 Classical models of urban economics

7.1.1 Why discuss these models here?

Urban and spatial economics are very active fields and produced several models of cities, their formation and evolution. As we will discuss later, these models are usually not completely validated by empirical data. They however give some hints and a theoretical framework for understanding some important stylized facts occurring in cities. These models usually rely on assumptions such as equilibrium, monocentricity, etc. that are not verified in most real-world sytems, but they identify relevant parameters and mechanisms. As such, they provide a basis for new approaches and models that integrate correctly the economic ingredients – but that try to avoid incorrect assumptions – and with predictions in agreement with data.

We believe that it is thus important to know and to understand these models in order to test them empirically, but also to be able to improve and to elaborate on them if needed. Since this literature is usually written by and for economists, we think that a simple explanation of the main results for non-economists could be helpful and valuable for anyone interested in cities and their economics. In particular, we will go through all the derivations in these models in a simple language accessible to scientists not trained in economics.

One of the first models was proposed by Von Thunen and provided an explanation of the spatial organization of rural areas around a central market where farmers sell their products. A later model that we will discuss in detail was proposed by different authors (now called the Alonso–Muth–Mills model), which considered individuals optimizing their utility and organizing themselves around a central business district where they work. We will also discuss the Beckmann model that introduces the social network in the utility. Later, Fujita and Ogawa considered a general version where both individuals and firms find their optimal location in the city. In particular they introduced an agglomeration effect by which it is beneficial for companies to be close to each other. This model which is a pillar of urban economics gives information about the spatial structure of cities and we discussed it in detail in Chapter 3 (Section 3.4).

7.1.2 The Alonso–Muth–Mills model

Most urban economics theory elaborate on the model proposed in the 1960s by Alonso, Muth, and Mills (see for example Fujita 1989 and references therein). This model relies on different assumptions. The first important one is the fact that all individuals behave in the same way and will all maximize the same utility function subject to the same budget constraint. The second assumption is the monocentric city structure organized around a unique central business district (CBD). The information about transport infrastructure is here absent and space is considered as homogeneous and isotropic. All individuals are assumed to work in this CBD and the transportation cost depends on the distance from home to it (here we can see the impact of the first model, proposed by Von Thunen (Von Thunen and Hall, 1966) with a unique market in an isolated state, where space is homogeneous and isotropic, and transportation cost depends only on the Euclidean distance to the central market).

The Alonso-Muth-Mills (AMM) model is defined by a utility function U that describes the preferences of individuals (or households – they are considered to be the same in this first simple approach). This function is usually assumed to depend on the land consumption s and on the composite commodity z (which corresponds to the money left when rent and transportation costs are substracted from the income)

$$U = U(z, s). \tag{7.1}$$

This utility has to satisfy the general constraints

$$\frac{\partial U}{\partial s} > 0, \qquad \frac{\partial U}{\partial z} > 0 \tag{7.2}$$

which just means that households in general prefer a larger apartment and smaller costs.

The budget constraint is given by

$$Y = z + T(x) + R(x)s \tag{7.3}$$

where Y is the income, $T(x)$ the transportation cost to work when living at location x, and $R(x)$ the renting cost per unit area at x. The problem is to optimize the utility subject to this budget constraint

$$\max_{z,s} U(z, s) \qquad \text{subject to} \qquad Y = z + T(x) + R(x)s. \tag{7.4}$$

In the following, we first discuss general results that can be obtained within this framework. We then discuss the bid-rent function that explicitly solves this problem and obtains rent and density profiles. Here and in the following we will consider essentially the one-dimensional case but most results can easily be extended to the more realistic 2d case.

General discussion

The maximization of the utility subject to a constraint is a classical problem which can be solved with Lagrange multipliers, but we will here discuss a faster way to obtain general results. We introduce the constraint with $z = Y - C_R - T$ and we maximize $U(Y - R(x)s - T(x), s)$ with respect to s

$$\frac{\partial U}{\partial s} = 0 = \partial_1 U(-R(x)) + \partial_2 U \tag{7.5}$$

where $\partial_i U$ denotes the derivative with respect to the ith variable. From this last equation, we obtain the renting cost under the form

$$R(x) = \frac{\partial_2 U}{\partial_1 U}. \tag{7.6}$$

An additional requirement is that the maximum utility U^* should be independent of x. If it is not, then individuals could choose another better location and we wouldn't be at equilibrium. We thus have to write

$$\frac{dU^*}{dx} = 0 = \partial_1 U \left(-s\frac{dR}{dx} - R\frac{ds}{dx} - T'(x) \right) + \partial_2 U \frac{\partial s}{\partial x} \tag{7.7}$$

where the functions $s(x)$ and $R(x)$ are computed at equilibrium ($T'(x)$ denotes dT/dx). Combining Eqs. (7.6) and (7.7), which are valid for all x, we then obtain the central result for the Alonso–Muth–Mills (AMM) model (see for example Brueckner 1987)

$$\frac{dR}{dx} = -\frac{T'(x)}{s(x)}. \tag{7.8}$$

This relation allows discussion of the location of individuals depending on their income and on their value of time (see Section 7.3). For example, for discussing the impact of income, we assume that transportation costs are linear in x implying $T'(x) = t$, and that we have two income categories, poor and rich, characterized by their (fixed) land consumption s_P and s_R, and transportation costs t_P and t_R. The category of individuals living in a given area of the city is then the one that is willing to pay more for the rent at this location. For example, for the poor living in the center we need the following condition to be satisfied (which is equivalent to the steepest bid-rent gradient condition)

$$\frac{t_P}{s_P} > \frac{t_R}{s_R}. \tag{7.9}$$

In the opposite case, rich individuals will live in the center, as they are willing to pay more than the poor for this location (details can be found for example in Fujita 1989).

The bid-rent function

We now discuss the general strategy to solve analytically these types of economic models. In general, we have a utility function U that describes the preferences of

7.1 Classical models of urban economics

individuals (or households). As above, we assume that this utility depends on s and z: $U = U(z, s)$, and the budget constraint is the same as above. Maximizing the utility under the budget constraint is in general difficult and economists devised a strategy that allows us to solve it in an easier way and which relies on the introduction of the "bid-rent function" (see for example Fujita 1989). The bid-rent function is defined as

$$\Psi(x, u) = \max_{z,s} \left\{ \frac{Y - T(x) - z}{s} \,\middle|\, U(z, s) = u \right\} \tag{7.10}$$

and represents the ability of a given household to pay for land under a fixed utility level. From $U(z, s) = u$ we can extract in principle the function $z = z(u, s)$, called the "indifference curve" (at utility level u). If we now take the derivative with respect to s we obtain

$$\frac{\partial}{\partial s} \frac{Y - T - z}{s} = 0 \tag{7.11}$$

and once this equation is solved we can replace s in its expression, leading to

$$-\frac{\partial z}{\partial s} = \Psi(x, u). \tag{7.12}$$

This last result means – for a given value of the utility u – that the tangent at the indifference curve is given by the bid-rent function.

We can now discuss the equilibrium location for a given household. We assume that $R(x)$ is an exogenous factor that describes the market land rent. We have to maximize the utility and this will be obtained when the bid-rent curve is a tangent to $R(x)$. This can be seen by introducing the so-called indirect utility function

$$V(R, I) = \max_{z,s} \{ U(z, s) | I = z + Rs \} \tag{7.13}$$

which gives the maximum utility attainable under the constraint of a net income I and land rent curve $R(x)$. This function can also be written as

$$V(R, I) = \max_{s} U(I - Rs, s). \tag{7.14}$$

If we set $R = \psi(x, u)$ and $I = Y - T(x)$, we then obtain

$$u = V(\Psi(x, u), Y - T(x)) \tag{7.15}$$

for all values of u. In other words the maximum utility under land rent $\Psi(x, u)$ and net income $Y - T(x)$ equals u. This relation is easy to understand, for we know that if we have a utility $U(z, s) = u$ the bid-rent function is a tangent to the indifference curve (Eq. (7.12)).

In addition, the maximum utility condition reads

$$\frac{dU}{ds} = 0 = \frac{\partial U}{\partial z}(-R) + \frac{\partial U}{\partial s} \qquad (7.16)$$

$$\Rightarrow R\frac{\partial U}{\partial z} = \frac{\partial U}{\partial s}. \qquad (7.17)$$

We now want to know how V varies with R and for this we compute

$$\frac{\partial V}{\partial R} = \frac{\partial U}{\partial z}\left(-R\frac{\partial s}{\partial R} - s\right) + \frac{\partial U}{\partial s}\frac{\partial s}{\partial R}$$

$$= -s\frac{\partial U}{\partial z} < 0. \qquad (7.18)$$

The indirect utility function is thus a decreasing function of R.

Now, if we denote by u^* the maximum utility that a household can achieve at an optimal location x^*, we have

$$u^* = V(R(x^*), Y - T(x^*)) \qquad (7.19)$$

$$u^* \geq V(R(x), Y - T(x)) \qquad (7.20)$$

which implies from Eq. (7.15) that

$$R(x^*) = \Psi(x^*, u^*) \qquad (7.21)$$

and since V is a decreasing function of R we also have $R(x) \geq \Psi(x, u^*)$.

These considerations imply that at equilibrium, the market rent is equal to the bid-rent function. In other words, each location is occupied by the individual who is able to offer the highest bid. This is the central result which allows us to tackle the problem of the equilibrium location of many households (of the same type). At equilibrium, all these households must achieve the same maximum utility level independent of the location. From this assumption, the rent curve is then given by

$$R(x) = \Psi(x, u^*) \text{ in the city} \qquad (7.22)$$

$$R(x) = R_A \text{ outside} \qquad (7.23)$$

where R_A is the value of the agricultural land rent.

There are different types of models and we discuss here the case of the closed city, where the population of the city is given (exogenous) and equal to P. This is in contrast with the open-city model, where households can move in or out of the city, determining the population endogenously. Since we are discussing here urban economics, we also specify the land ownership and the standard assumption is the

7.1 Classical models of urban economics

the one of absentee landlords model in which landlords are not living in the city and do not appear in this model.

In summary, the strategy for solving such a model is then given by the following steps:

- For a fixed value of the utility $U(z, s) = u$, find the indifference curve $z = z(u, s)$.
- Compute the bid-rent function by maximizing

$$\Psi = \frac{Y - T(x) - z(x, u)}{s} \qquad (7.24)$$

which implies that

$$-s\frac{dz}{ds} = Y - T - z. \qquad (7.25)$$

From this, we obtain the function $s = s(x, u^*)$ and $R(x) = \Psi(x, u^*)$.

- We are left with two unknowns, the city size b and the equilibrium utility u^*, which are determined by the constraint at the boundary and the closed-city condition. More precisely, if we consider a one-dimensional city, we have $R(x = \pm b) = R_A$ where R_A is the agricultural land rent at the fringe of the city, and (for a 1d city)

$$\int_{-b}^{b} \frac{dx}{s(x, u^*)} = P. \qquad (7.26)$$

We illustrate this strategy with the simple example of a two-dimensional monocentric city with business district located at $r = 0$. The utility is here given by the following function

$$U(z, s) = \alpha \log z + \beta \log s \qquad (7.27)$$

which can be normalized such that $\alpha + \beta = 1$. The budget constraint reads $Y = z + sR + T(r)$, where the transportation costs are assumed to be linear with the distance to the center $T(r) = ar$. In this case the indifference curve is given by

$$z(u, s) = e^{u/\alpha} s^{-\beta/\alpha}. \qquad (7.28)$$

In order to determine the bid-rent function $\Psi(u, r)$, we use Eq. (7.25), which reads

$$-se^{u/\alpha}\left(\frac{-\beta}{\alpha}\right)s^{-\beta/\alpha-1} = Y - T - e^{u/\alpha}s^{-\beta/\alpha} \qquad (7.29)$$

and leads to the following expression for the land consumption

$$s(r) = \alpha^{-\alpha/\beta} e^{u^*/\beta}[Y - ar]^{-\alpha/\beta}. \qquad (7.30)$$

We then determine the bid-rent curve, and the market rent curve is given by $R(r) = \Psi(r, u^*)$

$$R(r) = \beta \alpha^{\alpha/\beta} e^{-u^*/\beta}(Y - ar)^{1/\beta}. \qquad (7.31)$$

So far we thus have the two unknowns u^* and the size of the city r_f, and we now express the closed-city condition and the boundary condition

$$\int_0^{r_f} \frac{1}{s(r)} 2\pi r \, dr = P \tag{7.32}$$

$$R(r_f) = R_A. \tag{7.33}$$

If we assume that $R_A \approx 0$, we can easily solve this and we finally obtain for the rent curve and the density

$$R(r) = \frac{N\beta a^2}{2\pi Y^{2+\alpha/\beta} B(1, \alpha/\beta)} [Y - ar]^{1/\beta} \tag{7.34}$$

$$\rho(r) = [Y - ar]^{\alpha/\beta} \tag{7.35}$$

where $B(a, b)$ is the beta function. In this case we see that this simple model predicts a decrease of density with the distance to the center – which is in general expected – and equivalently an increase of the land consumption. The size of the city is here determined by the transport costs only, $r_f = Y/a$, which corresponds to the case where households convert completely their income in transportation cost.

As pointed out in Chapter 2, we see that the shape of all functions ($R(r)$, $\rho(r)$, etc) depends crucially on the choice of the utility $U(z, s)$. It is indeed a simple exercise to check that for example $U(z, s) = z + \alpha \log s$ would predict a density behaving as

$$\rho(r) \propto e^{-ar/\alpha} \tag{7.36}$$

which corresponds to the classical view of a monocentric city with exponentially decreasing densities (see for example Guérois and Pumain 2008 and references therein).

7.1.3 Beckmann's model: space and the social network

A model proposed some time ago by Beckmann (1976) meets the modern question of the coupling between the social network and the spatial location of individuals. The argument is that in the standard AMM model, work and consumption dominate all trip-making behavior and interaction with other residents through social contacts is neglected. In order to take into account these social interactions, Beckmann considers a standard utility that depends on the composite goods z and the residence area, of the form

$$\mathcal{U} = z + f(s) + S \tag{7.37}$$

where $f(s)$ is a function of the land space consumption s. This function should be increasing with s, signaling the preference for larger land space. In the original

7.1 Classical models of urban economics

paper Beckmann used $f(s) = \beta \log s$ and Mossay and Picard (2011) used the form $f(s) = -\beta/2s$ that we will keep in the following. The new term S represents the social utility for an agent interacting with all others and reads

$$S = B \int \rho(y) dy \qquad (7.38)$$

where $\rho(y)$ is the population density at point y. This term represents the benefit of interacting with all other individuals, and therefore it assumes no structure in the social network.

The budget constraint reads

$$Y = z + s R(s) + T(x) \qquad (7.39)$$

where $R(s)$ is the rent price per unit area and $T(x)$ is the cost for interacting in the social network. This cost can be written as

$$T(x) = A \int W(x-y) \rho(y) dy \qquad (7.40)$$

where Beckmann considered the particular case where the interaction cost for two individuals located at x and y is of the form $W(x-y) = t|x-y|$ (in the following we absorb A in t).

The advantage of this model is that we can solve it exactly, and we give below the outline of the derivation. The bid-rent function reads

$$\Psi = \max_s \frac{Y - T - z(u)}{s} \qquad (7.41)$$

which leads to

$$-s \frac{dz}{ds} = Y - T - z. \qquad (7.42)$$

The indifference curve $z = z(u)$ is given by

$$z = u + \frac{\beta}{2s} - S \qquad (7.43)$$

so that

$$\frac{\beta}{2s} = Y - T - u - \frac{\beta}{2s} + S \qquad (7.44)$$

from which we obtain

$$s(x) = \frac{\beta}{Y - u^* + S - T} \qquad (7.45)$$

$$R(x) = \frac{1}{2\beta}(Y - u^* + S - T)^2 \qquad (7.46)$$

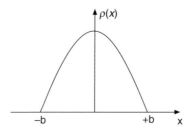

Figure 7.1. Population density obtained for the 1d Beckmann model. The size of the city is of order $b \sim \sqrt{\beta/t}$.

where u^* is the maximum utility. Here also we have two unknowns left at this stage: the maximum utility u^* and the city size b. We assume that rural land is characterized by a rent $R(\pm b) = 0$ and that the total number of individuals in the city is given by

$$\int_{-b}^{b} \rho(y) \, dy = P. \tag{7.47}$$

These two constraints allow us to determine the unknowns. The density $\rho(x) = 1/s(x)$ is then given Eq. (7.46) where T is itself a function of ρ as defined in Eq. (7.40). This integral equation on ρ reads

$$\rho(x) = \frac{C}{\beta} + \frac{1}{\beta} \int_{-b}^{b} (B - t|x - y|) \rho(y) \, dy. \tag{7.48}$$

By taking twice the derivative of this equation, we obtain the following differential equation

$$\frac{d^2 \rho}{dx^2} = -\frac{2t}{\beta} \rho(x) \tag{7.49}$$

whose solution is (Fig. 7.1)

$$\rho(x) = N \frac{\delta}{2} \cos \delta x \tag{7.50}$$

where the important scale is

$$\delta = \sqrt{\frac{2t}{\beta}}. \tag{7.51}$$

The quantity $1/\delta$ gives the typical size of the city (determined by $b\delta = \pi/2$). The size of the city decreases with the interaction cost, as expected. The interesting point here is that we observe within this model the spontaneous formation of a city center with a larger density. However, the assumption about the social network is incorrect as it assumes that all individuals interact with each other. It would be interesting to understand within this framework what the effect would be of a more realistic network structure on the spatial organization of a city.

7.2 Segregation and income structure of cities

Most cities are not homogeneous in space, and an important aspect concerns the spatial heterogeneity of social classes in cities. These distributions are not static and there is a whole dynamic over longer time scales, exemplified by gentrification (Atkinson and Bridge 2004; Venerandi et al. 2014) that reinforces the spatial heterogeneity of cities. The spatial distribution of income shapes the structure and organization of cities and its understanding has broad societal implications. It is therefore crucial to be able to characterize this spatial distribution, but despite an abundant literature in sociology but economics, there is no consensus on an adequate way to do this.

In many studies (usually concerning US cities), the question of the spatial pattern of segregation is limited to the study of the center versus suburbs and is usually adressed in two different ways. First, a central area is defined by arbitrary boundaries and measures are performed at the scale of this central area, while the rest is labeled as "suburbs." The issue with this approach is that the conclusions depend on the chosen boundaries and there is no unique, unambiguous definition of the city center: while some consider the Central Business District (Glaeser et al. 2008), others choose the urban core (urbanized area), where the population density is higher. The second approach, in an attempt to get rid of arbitrary boundaries, consists in plotting indicators of wealth as a function of distance to the center (Glaeser et al. 2008). This approach, inspired by the monocentric and isotropic city of classical economic studies, has however a serious flaw: cities are not isotropic and are spread unevenly in space, leading to very irregular shapes. Representing any quantity versus the distance to a center thus amounts to averages over very different areas and in polycentric cases is necessarily misleading.

Different aspects of segregation were characterized by a large number of indicators (see Massey and Denton 1988), and here, instead of attempting to enumerate these quantities, we define a null random model and observe in empirical data deviations from it. Segregation then appears as a significant deviation from a pure random case where individuals are distributed uniformly in the city (Louf and Barthelemy 2016). In addition, to allow for an objective characterization of segregation (once the null model is carefully defined), this method is sufficiently general and could be applied to a variety of other spatial phenomena.

In the next section, we thus define a null model for spatial segregation, and we focus on the empirical characterization of the income patterns in cities. In the next sections, we will discuss various theoretical approaches that aim at explaining these heterogeneities.

7.2.1 A null model for spatial segregation

When we deal with an empirical distribution, it is important to define a null model in order to understand where we stand and whether there is any significant deviation that would be the sign of something interesting happening. Here, we define such a null model for segregation, which then allows to characterize statistical deviations that may convey important information. We assume that the city has a total population N (we identify households and individuals) divided into different categories. In each category α we have N_α individuals in the city. From the spatial point of view, the urban area is divided into T aerial units, and in each unit t, we have $n(t)$ households divided into the different categories measured by $n_\alpha(t)$. We thus have the following constraints

$$\sum_\alpha n_\alpha(t) = n(t) \tag{7.52}$$

$$\sum_t n_\alpha(t) = N_\alpha \tag{7.53}$$

$$\sum_t n(t) = N, \tag{7.54}$$

We define a null model such that these constraints are satisfied and, in order to be comparable with the real city, $\{n(t)\}$ and $\{N_\alpha\}$ are the same for the null model and the real city. A natural choice for an unsegregated city is then obtained by distributing at random in the city the households of the different categories.

The problem of finding the numbers $(n_\alpha(1), \ldots, n_\alpha(T))$ of individuals belonging to a certain category α in the T areal units of an unsegregated city is a simple occupancy problem in combinatorics. If we assume that for all categories α, we have $n_\alpha(t) \ll n(t)$, we can neglect the constraint Eq. (7.52), keep Eq. (7.53), and we obtain the multinomial distribution

$$f(n_\alpha(1), \ldots, n_\alpha(T)) = \frac{N_\alpha!}{\prod_{t=1}^T n_\alpha(t)!} \prod_{t=1}^T p(t)^{n_\alpha(t)}, \tag{7.55}$$

where $p(t) = n(t)/N$ is the probability that a household randomly located belongs to the unit t. We then have for the average and the variance

$$\overline{n_\alpha(t)} = N_\alpha \frac{n(t)}{N}, \tag{7.56}$$

$$\mathrm{Var}[n_\alpha(t)] = N_\alpha \frac{n(t)}{N}\left(1 - \frac{n(t)}{N}\right). \tag{7.57}$$

In metropolitan areas, $N_\alpha \gg 1$ and the distribution of the $n_\alpha(t)$ can be approximated by a Gaussian with the average and variance given above.

7.2 Segregation and income structure of cities

The fundamental quantity that we will use in the following is the *representation* of a category α in the areal unit t, defined as

$$r_\alpha(t) = \frac{n_\alpha(t)/n(t)}{N_\alpha/N} = \frac{n_\alpha(t)/N_\alpha}{n(t)/N}. \tag{7.58}$$

The representation thus compares the relative population α in the areal unit t to the value that is expected in an unsegregated city where individuals in each class are randomly distributed according to their proportion. In the unsegregated case, we obtain

$$\overline{r_\alpha(t)} = 1,$$

$$\mathrm{Var}[r_\alpha(t)] = \sigma_\alpha(t)^2 = \frac{1}{N_\alpha}\left(\frac{N}{n(t)} - 1\right). \tag{7.59}$$

An important merit of the representation is the possibility of defining rigorously the notion of *over*-representation and *under*-representation of a population α in a geographical area. A population α is:

- over-represented in the geographical area t if $r_\alpha(t) > 1 + a\sigma_\alpha(t)$;
- under-represented in the geographical area t if $r_\alpha(t) < 1 - a\sigma_\alpha(t)$. For 99% confidence, we have to take $a = 2.57$.
- If the value $r_\alpha(t)$ falls in between the two previous limits, the representation of the population α is not statistically different (at this confidence level) from what would be obtained if individuals were distributed at random.

The representation allows to assess the significance of the deviation of population distributions from the unsegregated city. Also, it does not depend on the class structure at the city scale, but only on the spatial repartition of individuals belonging to each class. This is essential to be able to compare different cities where the group compositions – or inequality – might differ. Inequality and segregation are indeed two separate concepts, and should be measured differently.

7.2.2 The emergent social stratification of cities

Several difficulties are tied to the existence of different categories in data. In general, categories depend on a subjective definition that can vary in time and is usually different from one country to another. In particular, in the case of categories based on a continuum (such as income), the thresholds chosen to define classes are usually arbitrary (Jargowsky 1996). The representation defined above solves this issue by defining classes in an unambiguous and non-arbitrary way through their pattern of spatial interaction. We discuss here the case of the income distribution (using the data of the 2000 US Census at the Metropolitan Statistical scale), however, the method presented here is very general, and can be applied to different geographical levels, to an arbitrary number of population categories, and to different variables such as ethnicity, education level, and so on.

In order to construct categories, we first need a measure of the intensity of attraction (or repulsion) between different populations. The measure defined here is inspired by the M-value introduced by Marcon and Puech (2009) and used as a measure of interaction by Jensen (2006) – see also section 3.1.2. These authors were interested in measuring the geographic concentration of different types of industries. While previous measures (such as Ripley's K-value; see for example Dixon 2002) allow departures from a random (Poisson) distribution to be identified, the M-value's interest resides in the possibility of evaluating the tendency of different categories (such as firms) to co-locate. In the context of segregation, the idea is the following: we consider two categories α and β and we measure to what extent they are co-located in the same areal unit. We thus measure the average representation of the category β witnessed by individuals in category α, and obtain the following quantity $E_{\alpha\beta}$:

$$E_{\alpha\beta} = \frac{1}{N_\alpha} \sum_{t=1}^{T} n_\alpha(t) r_\beta(t). \qquad (7.60)$$

We note that this measure is symmetric, $E_{\alpha\beta} = E_{\beta\alpha}$, as can be checked directly on the expression.

This "E-value" is a measure of exposure according to the typology of segregation measures proposed by Massey and Denton (1988). It is however different from the traditional measure of exposure and allows attraction ($E > 1$) or repulsion ($E < 1$) to be characterized between categories. In the case of an unsegregated city, every household in α sees on average $r_\beta = 1$ and we have $E_{\alpha\beta} = 1$. If populations α and β attract one another, they tend to be over-represented in the same areal units, and every household α sees $r_\beta > 1$, leading to $E_{\alpha\beta} > 1$. On the other hand, if they repel each other, every household α sees $r_\beta < 1$ and we have $E_{\alpha\beta} < 1$. The minimum of exposure for two classes α and β is obtained when these two categories are never present together in the same areal unit and then

$$E_{\alpha\beta}^{min} = 0. \qquad (7.61)$$

The maximum is obtained when the two classes are alone in the system and we get

$$E_{\alpha\beta}^{max} = \frac{N^2}{4N_\alpha N_\beta}. \qquad (7.62)$$

In the case, $\alpha = \beta$, the E-measure represents the "isolation," defined as

$$I_\alpha = \frac{1}{N_\alpha} \sum_{t=1}^{T} n_\alpha(t) r_\alpha(t), \qquad (7.63)$$

and measures to what extent individuals from the same category interact with each other. In an unsegregated city, where individuals are indifferent to others when choosing their residence, we have $I_\alpha^{min} = 1$. In contrast, in the extreme situation where individuals belonging to the class α live in isolation from the others, the isolation reaches its maximum value

$$I_\alpha^{max} = \frac{N}{N_\alpha}. \qquad (7.64)$$

In order to discuss the significance of the values of exposure and isolation, one needs to compute the variance of the exposure in the unsegregated situation defined earlier (other details and calculations for the variance as well as for the extrema are presented in Louf and Barthelemy 2016).

Empirical results

The emergent social stratification of cities. The income subdivision operated by national agencies does not usually rely on a scientific basis but is essentially defined for practical reasons. Grouping individuals according to such classes could be misleading and it would be more satisfying to see such classes emerging from the raw data itself. In order to do that, we assume that social classes are reflected in the behavior of individuals: households belonging to the same class tend to live together, while households belonging to different classes tend to avoid one another. The idea is thus to define classes based on the way they manifest themselves through the spatial repartition of the different categories. Starting from the finest income subdivision given by the census bureau (16 subdivisions), we then compute the 16 × 16 matrix of $E_{\alpha\beta}$ values for all cities, and perform a hierarchical clustering on this matrix, successively aggregating the subdivisions with the highest $E_{\alpha\beta}$ values. We stop the aggregation process when the only classes left are indifferent ($E_{\alpha\beta} = 1$ with 99% confidence) or repel each other ($E_{\alpha\beta} < 1$ with 99% confidence) (for more details on the aggregation procedure, see Louf and Barthelemy 2016). Strikingly, the outcome of this method is the emergence of three distinct classes (see Fig. 7.2):

- The higher-income (>45,000USD), which represents 47% of the US population.
- The lower-income (<35,000USD) classes (42% of the US population). The high- and low-income classes repel each other strongly while being respectively very coherent.
- A small middle-income (35,000–45,000USD) class (11% of the population) that is relatively indifferent to the other classes.

This result implies that there is some truth in the conventional way of dividing populations into three income classes, and what we casually perceive as the social

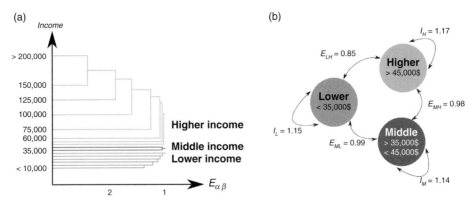

Figure 7.2. (a) Alluvial diagram showing the successive aggregations of different income categories in the clustering process, and the value of the exposure at which the aggregation took place. The aggregation stops when there is no pair of categories for which $E > 1$, when all classes are at best indifferent to one another. The highest-income categories attract each other with more intensity (higher values of $E_{\alpha\beta}$) than the lowest-income categories. (b) The emerging classes and their respective exposure and isolation values. The lower- and higher-income classes repel each other, while the middle-income class is indifferent to the other classes. The higher-income class is more coherent than the lower-income, which is more coherent than the middle-income class, as reflected by the isolation coefficient I. Figure taken from Louf and Barthelemy 2016.

stratification in our cities actually also emerges from the spatial interaction of people.

Larger cities are richer. At the scale of an entire country, segregation can manifest itself in the unequal representation of the different income classes across urban areas. We plot on Fig. 7.3 the ratio

$$\frac{N_\alpha^>(H)}{N^>(H)} \tag{7.65}$$

where $N^>(H)$ is the number of cities of population greater than H, and $N_\alpha^>(H)$ the number of cities of population greater than H for which the class α is over-represented (ie., $\overline{r_\alpha} > 1$). When plotting this ratio versus H, a decreasing curve indicates that the category α tends to be under-represented in larger urban areas, while an increasing curve shows that the category α tends to be over-represented in larger urban areas (the representation here is measured with respect to the total population at the US level). We observe that the number of cities where higher-income households are over-represented increases with the size of the cities, while the inverse trend is true for lower-income households.

Poor center, rich suburbs? The classical approach to this problem in economics is to measure quantities averaged at a certain distance from the center of the

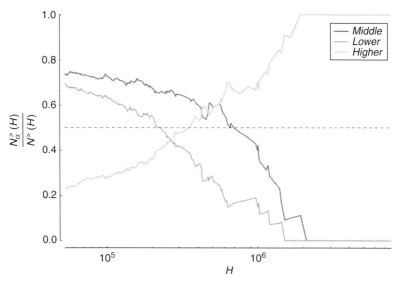

Figure 7.3. Proportion of cities in which the different classes are over-represented, as a function of their total population. As cities get larger, rich people are over-represented and poor people under-represented (compared to national levels). Figure taken from Louf and Barthelemy (2016).

city (see for example Glaeser et al. 2008). For monocentric, isotropic cities this is indeed justified, but as soon as cities become larger their structure becomes complex, with multiple centers. These cities are not monocentric anymore and averaging a quantity over a circle at a distance from a certain center has no meaning and mixes many different structures. We discuss here a different approach that does not require the definition of a distance to the center. Instead, we plot the average representation computed over all area units (census-blocks in the US case) with a density population in a given interval $[\rho, \rho + d\rho]$. This sheds some light on the difference of social composition between the high-density and low-density areas in cities. As shown in Fig. 7.4, low-density regions in cities are on average rich neighborhoods, although the effect is less pronounced than in the center. There is also a surprising result: areas with very large densities (typically above 20,000 inhabitants/km^2) are on average rich neighborhoods, and slightly lower densities are on average poor neighborhoods. Only a few cities in the US have neighborhoods that reach this threshold of 20,000 inhabitants per km^2, which can explain why we observe in most cases poor centers and rich suburbs. This result suggests that density could be important and relevant in the usual discussion about income structure differences between North American and European cities.

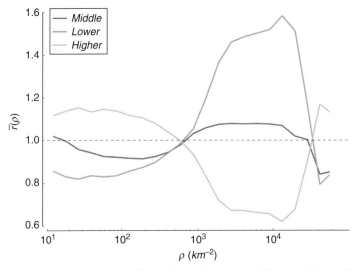

Figure 7.4. Average representation of the higher-, middle- and lower-income classes over the 276 MSA in the US as a function of the local density of households. On average, we find that low-density areas (the suburbs) are rich, while high density regions (the center) are poor, confirming empirically on a large dataset stylized facts that had previously emerged from local studies. Interestingly, very large-density areas ($\rho > 20{,}000 \text{ km}^{-2}$) are rich on average, suggesting that density may be one relevant factor in explaining the differences between neighborhoods. Figures taken from Louf and Barthelemy 2016.

7.3 Modeling segregation

7.3.1 Transportation modes in the Alonso–Muth–Mills model

An important empirical fact discussed in the previous section is that in the US, richer people tend to live in the suburbs, while the poor live in the center of cities (it is usually the opposite in Europe and we discuss this at the end of this section). This important fact should then find some explanation, and for economists it is natural to start from the Alonso–Muth–Mills model (see Section 7.1) and to see how this effect could be explained within this framework. This is what Glaeser et al. (2008) did and since it is an important aspect of cities, we reproduce here the main arguments of this work. They considered a simple version of the AMM model with two categories of people: rich and poor, characterized by their income $Y_{rich} > Y_{poor}$, land consumption $s_{rich} > s_{poor}$ (supposed to be given for each category), and values of time V_{rich} and V_{poor}.

We first discuss here the value of time. This quantity is an important theoretical ingredient that allows understanding of how individuals choose one transportation mode over another. The main point is to introduce the generalized cost $C(m, i, j)$ of a trip between locations i and j with mode m, which is given by the sum of

7.3 Modeling segregation

financial costs C_{fin} and non-monetary cost given by the time spent $\tau(m; i, j)$ multiplied by the value of time V

$$C(m, i, j) = C_{fin} + V\tau(m; i, j). \qquad (7.66)$$

The prefactor V (in money per unit time) thus converts the time spent on a trip into a monetary cost. An individual with a large value of time will tend to prefer rapid transportation modes even if they have a high financial cost. Although, it is difficult to estimate empirically, it is usually assumed that the value of time increases with income.

For example, for a distance d, we write for cars (C_1) and for the subway (C_2)

$$C_1(d) = C + V\frac{d}{v_c} \qquad (7.67)$$

$$C_2(d) = \left(F + \frac{d}{v_s}\right)V \qquad (7.68)$$

where C is the financial cost of a car, F the additional time for subway transportation, and v_c and v_s the speed of cars and subway, respectively. For a given distance d, these expressions then allow us to determine which mode will be used: mode 2 will be preferred over mode 1 with a probability that depends on the difference $p = \mathcal{F}(C_1 - C_2)$, where the function $\mathcal{F}$ is usually assumed to be the logit function

$$\mathcal{F}(\Delta C) = \frac{1}{1 + e^{-\Delta C}}. \qquad (7.69)$$

For $\Delta C = C_1 - C_2 \to +\infty$ the probability of choosing 2 is one (and zero for $\Delta C \to -\infty$). This simple choice is consistent with the fact that when the difference of costs is large, the choice is easier to make.

Glaeser et al. (2008) assumed that there are essentially two transportation modes, characterized by different velocities v_s, v_c (slow and fast modes such as subway and car). For each income group i, in the case of a monocentric (and isotropic) city, for a transport mode m, the utility maximization leads to the following equation for the bid-rent function R_i (see Eq. (7.8)),

$$\frac{\partial R_i}{\partial r} = -\frac{V_i}{v_m s_i} \qquad (7.70)$$

where we assumed that for a given transportation mode of velocity v_m, the cost is $T_m(r) = V_i r / v_m$, where V_i is the value of time of income group i and s_i its land consumption. The steepness of this bid-rent function will determine which group lives closest to the city center. Indeed, the group with the steepest gradient is willing to pay more than the other group for living in the city center. For a given transport mode, the poor then live in the center if

$$\frac{S_{rich}}{S_{poor}} > \frac{V_{rich}}{V_{poor}}. \tag{7.71}$$

This relation can be rewritten in terms of the "income elasticity of demand for land," defined as

$$\varepsilon_Y^s = \frac{Y_{poor}}{S_{poor}} \frac{S_{rich} - S_{poor}}{Y_{rich} - Y_{poor}} \tag{7.72}$$

and the income elasticity of time cost

$$\varepsilon_Y^V = \frac{Y_{poor}}{V_{poor}} \frac{V_{rich} - V_{poor}}{Y_{rich} - Y_{poor}} \tag{7.73}$$

and reads

$$\varepsilon_Y^s > \varepsilon_Y^V. \tag{7.74}$$

The poor thus live in the city center when the income elasticity for land is larger than the income elasticity of time cost. In other words, this happens when the preference for land increases with income faster than the cost of time. According to empirical estimates discussed by Glaeser et al. (2008), it seems that in the US this inequality is not satisfied, and the model, when only one transportation mode is taken into account would thus predict that rich individuals are living in city centers in contradiction with empirical facts.

In order to solve this puzzle, we assume with Glaeser et al. that there are two different transportation modes with different velocities and costs. In a simple case, they assume that both income groups take public transportation. If $V_R F > C > V_P F$, then members of the poorer group will take public transportation for distances such that $C_2(d) < C_1(D)$, leading to (see Eqs. (7.67,7.68))

$$d < d^* = \frac{C - FV_P}{V_P(\frac{1}{v_s} - \frac{1}{v_P})}. \tag{7.75}$$

Within these conditions, the poor will then live in the center if the steepest bid-rent gradient condition is met

$$\frac{S_{rich}}{S_{poor}} > \frac{v_s}{v_c} \frac{V_{rich}}{V_{poor}}. \tag{7.76}$$

In this study, the ratio v_s/v_c is smaller than one, and this condition is therefore more likely, met, even if $\varepsilon_Y^s < \varepsilon_Y^V$. This simple argument shows that the poor have a comparative advantage in using the subway, and furthermore that public transportation has a comparative advantage for commuting short distances from the center. Public transportation increases here the probability that the poor will live in the center, in agreement with empirical observations for the US.

At this stage there are however many questions left. First, how can we understand the fact that the pattern in European cities is exactly the opposite, with rich people living in the center of cities, even in the presence of a dense public transportation network in the center? Brueckner et al. (1999) proposed an explanation of this phenomenon with the help of amenities. They extended the utility of the AMM model by introducing an exogenous amenity level at distance x from the center and given by an unknown function $a(x)$. The utility is then a function of z, s, and also of $a(x)$

$$u = u(z, s, \{a(x)\}). \tag{7.77}$$

The budget constraint is still $Y = z + Rs + Vx/v$, where V and v are the value of time and velocity of a given transportation mode. Repeating the calculation above, we easily find that

$$\frac{dR}{dx} = -\frac{V}{vs(x)} + \frac{u^a}{s(x)u^e} a'(x) \tag{7.78}$$

where $u^{a(e)} = \partial u/\partial a(e)$. This expression helps us to qualitatively understand the effect of amenities. In particular, by considering two income groups (poor and rich), Brueckner et al. (1999) showed that when the center has a strong amenity advantage over the suburbs, rich individuals will live in the center, which is what we observe in Europe. In the opposite case (such as in the US), the rich live in the suburbs. While this theory is elegant and predicts correctly the existence of a large variety of city structures, it relies on a utility function describing the amenities which seems very difficult to assess quantitatively. In some sense, the result is not surprising as the complex effect of amenities is somehow encoded in the input of the model under the form of an unknown utility. From an empirical point of view these approaches are not thoroughly tested, and one of the future challenges for a science of cities is to find a simple model with clear and testable assumptions and predictions.

7.3.2 A simple model for tie formation

The rank defined as (Liben-Nowell et al. 2005)

$$\text{rank}(i, j) = \{\#\text{nodes } w \text{ such that } d(i, w) < d(i, j)\} \tag{7.79}$$

is an important parameter that describes social interactions (see also Section 5.2, where this parameter appears as a crucial ingredient in mobility). It takes into account the fact that for low-density areas the distance is less important and people even far away are likely to know each other, which is not the case for very dense areas. It is then natural to write that the probability that two individuals located at

i and j know each other is (Liben-Nowell et al. 2005)

$$P_{ij} \sim \frac{1}{\text{rank}(i,j)} \tag{7.80}$$

This relation was verified empirically for a social network (Liben-Nowell et al. 2005) and in another study, Noulas et al. (2012) confirmed the importance of the rank in studying intra-urban movements. If we accept this form for the probability of being friends, we can consider the result for the uniform distribution with density ρ. The rank then reads $\rho\pi r^2$, where r is the distance from a given node, and the probability of having a friend at distance r then reads

$$P(r) = \frac{1}{\rho\pi r^2}. \tag{7.81}$$

The total number of ties for a given individual is then (Pan et al. 2013)

$$t(\rho) = \int_{r_{min}}^{r_{max}} 2\pi r \, dr \rho P(r) \tag{7.82}$$

where $r_{min} = 1/\sqrt{\pi\rho}$ is the minimal distance between individuals and r_{max} is the "boundary" of the city. Pan et al. considered that r_{max} is independent of the density (a disputable assumption, as we know that the area depends on the population) and we then find that the total number of ties T scales with density as

$$T = \rho \log \rho + C\rho \tag{7.83}$$

where C is a constant. This result was tested on of the spread of HIV and GDP, for example (Pan et al. 2013). The fits for T versus ρ are actually as good as a linear one or a power-law fit with exponents whose value is close to one (of the form $1+\epsilon$, where epsilon will depend on the noise and can be of order 0.1–0.2): it is thus, a "weak" proof. This reinforces the idea that scaling exponents very close to one are probably not enough to provide a good test for theories, and we should look for more robust ones.

7.3.3 Statistical physics of the Schelling model

Schelling (1971) proposed a simple model for understanding the dynamics of segregation. Agents are divided into two species that can move on a checkerboard according to a given utility function. Schelling showed that within this model, even when individuals have a mild preference for their own species, the system evolves toward a segregated state. However the physics of this model was only clarified recently in a numerical sudy (Dall'Asta et al. 2008). In this work, the authors consider the model defined in a previous study (Vinković and Kirman 2006), where

7.3 Modeling segregation

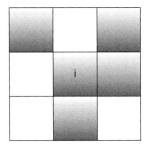

Figure 7.5. On this two-dimensional lattice (with a Moore neighborhood), the central gray site is surrounded by 8 sites, 4 whites and 4 gray. The fraction of neighbors of a different type is then $f = 4/8 = 1/2$ and according to the value chosen for the threshold f^*, this individual will move or stay at his current location.

individuals are of two types (± 1) and are initially randomly distributed on a line ($d = 1$) or on a regular lattice ($d = 2$) of $N = L^d$ sites. Each site i is described by a spin-like variable σ_i which takes the value $\sigma_i = 0$ if the site is empty, or $\sigma_i = \pm 1$ if the site is occupied by an individual of type ± 1. For a given site, we denote by f the fraction of its neighboring sites of a different type, and the dynamical rule is the following one: if f is less than some threshold f^* (taken equal to $1/2$ as in the original Schelling model), the individual has utility 1 and stays at this location. Otherwise, the individual is unhappy in this location (utility 0) and there are essentially two types of dynamics (Dall'Asta et al. 2008):

- The *constrained dynamics* is obtained when unhappy individuals are allowed to move (and only if they can find a vacancy in which they are happy).
- The *unconstrained dynamics* is where all agents are allowed to move as long as their situation does not get worse.

For both dynamics, the movers are allowed to visit any vacancy with no spatial constraints (which is called long-range diffusion in physics terms).

In magnetic language, the global magnetization $\sum_i \sigma_i$ is constant during these dynamics, but obviously the local magnetization can change. In order to characterize the state of the system, we consider the density of unhappy sites $u(t)$ and the density of interfaces $n(t)$, which measures the fraction of neighboring spins that are of opposite signs. The important quantity is thus $u(\infty)$, which is the remaining fraction of unhappy individuals after a long time. We denote by $\rho_0 = N_0/N$ and $\rho_\pm = N_\pm/N$ the densities of vacancies and of occupied sites of a given type.

In the constrained case, an individual is drawn at random and moved to a vacancy taken at random if it increases its utility (if there is no such vacancy the individual does not move). For this dynamics, the system converges to a Nash equilibrium where no individual can improve his utility by a single move (even if a

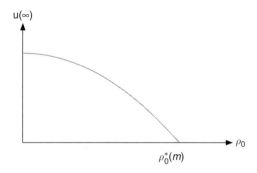

Figure 7.6. Density of unhappy spins (schematic). For a magnetization $m \neq 0$, there is a critical density $\rho_0^*(m)$ below which a non-zero fraction of unhappy individuals remains at long times (Dall'Asta et al. 2008).

better situation exists, it cannot be reached by single moves but only collectively). Starting from random initial conditions, numerical simulations (Dall'Asta et al. 2008) reveal the existence of a continuous transition according to the global magnetization $m = \rho_+ - \rho_-$ and the vacancy fraction ρ_0 (see Fig. 7.6). For $m \neq 0$, there is a threshold $\rho_0^*(m)$ above which $u(\infty) = 0$. Below this threshold, a finite fraction of unhappy individuals remains. When $m \to 0$, this threshold ρ_0^* decreases towards 0.

For the unconstrained dynamics, in two dimensions (which is the most interesting here), we expect that voids are distributed randomly in the system. Indeed, if the vacancy density is too large a cluster of void sites percolates through the system and prevents the growth of spin clusters, which are then limited to the typical size of clusters for this void density ρ_0. If ρ_0 is below the percolation threshold, the system converges to a quasi ordered state with two main domains spanning the whole system. The typical length of clusters grows as $t^{1/z}$, with $z = 2$ in agreement with general arguments about spin dynamics (Sen 1999). The behavior of the system with the threshold f^* is also very interesting: for $\rho_0 \to 0$, the coarsening process is robust for a wide range of f^* around $1/2$, and indeed segregation takes place even for large values such as $f^* \approx 5/8 = 0.625$. Above the value $f^* \geq 6/8$, no segregation takes place and the system remains in a disordered state. Maybe surprisingly, no coarsening takes place for extremely intolerant individuals: for $f^* = 1/8$, the system remains trapped in a disordered state and cannot reach the optimal fully segregated state. This behavior is confirmed in the case of a variant of the Schelling model (Gauvin et al. 2009), where the agents are satisfied if the number of unlike agents is lower than a fraction of all agents present in the neighborhood. In this case too, Schelling's intuition is correct: there is a large range of parameters for which the system converges to a segregated state.

7.3.4 Collective versus individual dynamics

In the Schelling model (and most of its variants), there is no collective optimization: each individual moves in order to improve its own utility, irrespective of the state of his neighbors. An individual move can then cause a larger number of unhappy sites. The fact that economical agents maximize their own utility is in contrast with what usually happens in physics, where systems of particles are in a state that optimizes a global quantity (such as energy). In order to understand the link between microscopic ingredients and collective behavior, Grauwin et al. (2009) proposed a Schelling-like model which interpolates continuously between individual and cooperative dynamics. In this model, they showed that individual maximization fails to lead the system to the social optimum.

We discuss the model for one type of agents (the extension to different types does not present any difficulty; see Grauwin et al. 2009). The city is divided into Q blocks ($Q \gg 1$), and each block contains H apartments that can contain at most one agent (or household). The number of agents in a given block i is $n_i \leq H$ and the local density is $\rho_i = n_i/H$. Each agent has the same utility function $u(\rho_i)$ which describes the preference of the agent with respect to the local density. To explicitly obtain the equilibrium configurations, Grauwin et al. choose the following specific form for the utility function u

$$u(\rho) = \begin{cases} 2\rho & \text{if } \rho \leq 1/2 \\ m + 2(1-m)(1-\rho) & \text{if } \rho > 1/2 \end{cases} \qquad (7.84)$$

where $m \in [0, 1]$. For this choice agents prefer half-filled ($\rho = 1/2$) or overcrowded neighborhoods ($\rho = 1$) compared to empty ones with $\rho = 0$.

The collective utility defined as the total utility of all agents is then

$$U(\{\rho\}) = H \sum_i \rho_i u(\rho_i) \qquad (7.85)$$

which depends on the configuration (denoted by $\{\rho\}$) of local densities in the city. At each time step, an agent taken at random will move to a vacant apartment according to the probability

$$P(\{\rho\} \to \{\rho'\}) = \frac{1}{1 + e^{-G/T}} \qquad (7.86)$$

where $\{\rho\}$ and $\{\rho'\}$ are the configurations before and after the move. The parameter T is the "temperature" that introduces noise in the decision process (if $T = 0$ the process is deterministic), and the quantity G is the gain associated with this move and is chosen to be

$$G = \Delta u + \alpha(\Delta U - \Delta u). \qquad (7.87)$$

The quantity Δu is the variation of the agent's utility and ΔU is the variation of the collective utility, both calculated for this specific move. The parameter $\alpha \in [0, 1]$ is thus a measure of the degree of cooperation between individuals. For $\alpha = 0$, the probability depends on the agent's interest only (in the spirit of Schelling's model), while for $\alpha = 1$ the decision to move depends on the collective advantage only. This parameter α can thus be seen as the existence of incentives created by a central government, for example.

In the case where the gain can be written as $G = V(\rho') - V(\rho)$, the detailed balance is satisfied and the stationary probability distribution $\Pi(\rho)$ is given by

$$\Pi(\rho) = \frac{1}{Z} e^{F(\rho)/T} \tag{7.88}$$

where $F(\rho) = V(\rho) + TS(\rho)$ (Z is the normalization constant, and the entropy S is a function of ρ that can be easily calculated). If we cannot construct such a function V, then the detailed balance is not satisfied and there are no general results so far. It doesn't mean that this case is not relevant, and in fact it might actually be the opposite, with the existence of local equilibrium currents that would be interesting to understand in a socio-economical context (Bouchaud 2013).

We can now discuss the equilibrium configurations, and for the sake of simplicity we consider that $T \to 0$ and we consider the case where the overall city density is given by $\rho_0 = 1/2$. This is a particularly interesting case: in principle there is a global social optimum for which all individuals are happy ($\rho_i = 1/2$ maximizes all local utilities). However, we observe the following results, shown in Fig. 7.7. In the collective case, $\alpha = 1$, the equilibrium corresponds to the maximum of the collective utility, obtained for $\rho_i = 1/2$ for all i. In contrast, for the "selfish" case, $\alpha = 0$, the maximization leads to segregation (Fig. 7.7, right). In this case, a fraction of the blocks are empty and the others have a density larger than $1/2$. This is obviously not an optimal situation, but it is a Nash equilibrium where no single move can improve the individual's utility. More precisely, a detailed calculation (Bouchaud 2013) shows that the stationary equilibrium corresponds to complete segregation ($\rho = 0$ or $\rho = 1$).

There is therefore a transition from a mixed to a segregated state when α goes from 1 to 0, or in other words, when the cooperation decreases (or the influence of government incentives diminishes). This transition is driven by the competition between the collective and the individual components of the dynamics, and with the choice in Eq. (7.84), Grauwin et al. (2009) computed the critical value $\alpha_c = 1/(3 - 2m)$. As noted by Bouchaud (2013), this result is particularly interesting and in sharp contrast with the "invisible hand" theory of Adam Smith. According to this theory, individual actions can result in unintended social benefits, and here we observe a clear counterexample where the individual optimization can actually

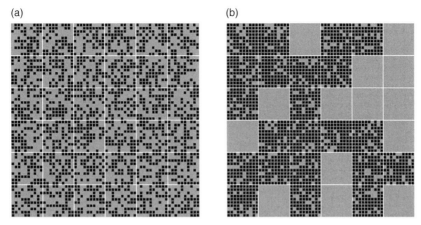

Figure 7.7. Equilibrium configurations at $T \to 0$, for a city composed of 36 blocks each containing $H = 100$ apartments, with $\rho_0 = 1/2$. Black cells are occupied and gray cells represent empty sites. (a) Collective optimization obtained for $\alpha = 1$; (b) segregation configuration obtained for $\alpha = 0$. Figure taken from Grauwin et al. (2009).

lead to a state very far from the social optimum, even when the global incentive exists ($\alpha < \alpha_c$).

Obviously, this is certainly not the end of the story, but this example shows how the tools and concepts borrowed from statistical physics can bring valuable and new insights about the dynamics of social systems.

7.4 Scaling in urban systems

The recent craze for scaling laws (Batty 2008; Bettencourt et al. 2007; Pumain 2004) has been an important step in the study of urban systems. We discuss here what is meant by scaling, and we then discuss a theoretical approach proposed recently (Bettencourt 2013).

7.4.1 What is scaling?

How various properties of a system vary with its size is an important question that has been addressed for a long time in physics. This notion of scaling was at the heart of statistical physics in the twentieth century and led to many important discoveries and a deep understanding of phenomena such as phase transitions (see for example Goldenfeld 1992). Scaling laws are not present only in physics, but can naturally be found in many different fields such as biology and mathematics. They are not a particular case of general relations, but indicate something important that never happens by accident (Barenblatt 1996).

In the case of urban systems, we can measure a number of global parameters for a given definition of a city. Area and population are the simplest ones, but we can think of other measures that convey social or economic information, such as the number of patents, the number of highly educated individuals, the number of crimes, etc. An interesting question discussed by Pumain (2004) concerns the evolution of a given parameter Y (usually extensive) with the population P. For many quantities, it seems that we observe a scaling behavior of the form

$$Y \sim P^\beta \tag{7.89}$$

where the exponent β is usually positive. This relation implies that if the population is multiplied by a factor λ, the quantity Y is multiplied by a factor less than or larger than λ. When $\beta > 1$ for example, its production is then "catalyzed" by the number of individuals. This type of scaling relation is a signature of various processes governing the phenomenon under study, especially when the exponent β is not what is naïvely expected (Barenblatt 1996). We note that this scaling law, Eq. (7.89), has not to be confused with the long-tail behavior for a probability distribution of the form $P(x) \sim x^{-\gamma}$, as observed for example for the income by Pareto.

As noted by Thisse (2014), such a scaling implies that cities are scaled versions of each other, but in fact only as far as the quantity Y is concerned. Probably not all quantities satisfy a scaling relation and, globally speaking, cities are not scaled versions of each other. More importantly, the existence of scaling means that there is a mechanism at the heart of cities that governs the evolution of Y. Also in this vision of a city, the population is the explanatory variable and within this framework it is essentially the clock governing the evolution of cities. Ideally, the population would be an endogenous variable but at this stage of our understanding of cities, a full model seems rather out of reach.

From Eq. (7.89), we see that the quantity per capita then scales as

$$\frac{Y}{P} \sim P^{\beta-1}. \tag{7.90}$$

This last equation shows that we can classify parameters in three large categories according to the value of β compared to one (see Bettencourt et al. 2007 and various examples in Table 7.1):

- The first, "trivial" case is $\beta = 1$, which implies that the quantity per capita Y/P is constant. This is the case for example with water consumption: it is a human-related quantity and does not depend on the size of the city.
- The case $\beta < 1$ is essentially associated with an economy of scale. This is the case for infrastructure quantities such as the total length of electric cables, etc.

7.4 Scaling in urban systems

Table 7.1. *Example of quantities displaying either sublinear or superlinear behavior.*

Quantity	Exponent β
Road length	1/2
Total commuting distance	0.6
Gasoline stations	0.77*
Length of electric cables	0.87*
Number of patents	1.27*
GDP	1.26*
Number of crimes	1.16*
Q_{gas,CO_2}	1.22–1.38†

*Bettencourt et al. (2007); †Oliveira et al. (2014).

- Finally, the case $\beta > 1$ corresponds to the case where the quantity per capita increases with the city size. A simple example is the total number of possible relations between individuals, which is $P(P-1)/2$ and implies for large P a value $\beta = 2$. This case is usually associated with the positive impact of cities.

The exponent β can sometimes be extracted from dimensional analysis, as physical laws cannot depend on an arbitrary choice of measurement units. We refer the reader interested in dimensional analysis to the book by Barenblatt (1996), and we just sketch here the simple main idea. If we have a quantity Y which depends on a set of variables X through the equation $Y = F(X)$, then the dimension (the unit) of Y should be exactly the one of F. For example, if we have a length L, a mass m and a density ρ, the only way to express L in terms of m and ρ is

$$L = \left(\frac{m}{\rho}\right)^{1/3}. \tag{7.91}$$

If there is another length L', the exact relation cannot be determined by dimensional considerations anymore and we can just write a relation of the form

$$Lf\left(\frac{L}{L'}\right) = \left(\frac{m}{\rho}\right)^{1/3} \tag{7.92}$$

where f is an unknown function at this stage. More subtle results can be obtained with dimensional analysis, but in a variety of interesting problems, exponents cannot be obtained through these simple arguments. This usually happens when there are some parameters (length scales, for example) that escaped from our first analysis. For example, if we have a scaling relation of the form $Y = AX^\beta$, with quantities Y and X having different dimensions and $\beta \neq 1$, it can be rewritten as

$$Y = Y_0 \left(\frac{X}{X_0} \right)^{\beta} \tag{7.93}$$

where X_0 (resp. Y_0) has the dimension of X (resp. Y). We thus see here that two quantities X_0 and Y_0 need to be introduced and represent new scales associated with the process under study.

The values of these exponents β thus provide a very useful guide for theoretical analysis: a simple model able to explain the values of these exponents allows identification of the most relevant mechanisms. This was the goal of the proposal by Bettencourt (2013), where a relatively small number of assumptions should be able to reproduce the nonlinearities observed empirically (see next section).

We note the striking fact that many exponents reported so far in the literature are actually very close to one. We have then to carefully check the effect of noise on the value of the exponent and the possibility that there are some errors in the fitting process (Leitao et al, 2016). In addition, we note also that many of these quantities are actually broadly distributed in the city, casting some doubt on the validity and the meaning of the corresponding scaling. These problems are particularly relevant for cities where the value of these exponents varies with the definition of the city, as shown by Arcaute et al. 2013 (see also Section 2.5 about CO_2 emissions). This could be the sign of serious fitting problems or of other problems related to large fluctuations in cities.

7.4.2 Theoretical approaches

Non-trivial values of exponents observed in scaling laws have triggered the interest of many scientists, including statistical physicists. Indeed, for many systems the fact that an exponent is not given by its "trivial" value is usually a sign that something interesting is going on. At least it means that there are other spatial and time scales, scales that are in turn related to some unknown mechanism.

These exponents can then serve as a guide for theoretical approaches: any reasonable model should be able to explain their value, or at least, give some approximate values obtained under minimal assumptions (usually called mean-field models in physics).

In the introduction (Chapter 1), we discussed naïve calculations of the exponents and in Chapter 3, we presented an approach based on the Fujita–Ogawa model. This simple model allows explanation of the scaling of the number of polycenters and other mobility indicators. In particular, we found that for the number of activity centers h, the total length of roads L_N, the total commuting

7.4 Scaling in urban systems

length L_{tot}, the area A, the scalings are of the form

$$h \sim P^{\beta_h} \tag{7.94}$$

$$L_N \sim P^{\beta_N} \tag{7.95}$$

$$L_{tot} \sim P^{\beta_{tot}} \tag{7.96}$$

$$A \sim P^{\beta_A} \tag{7.97}$$

where in this theoretical approach $\beta_H < 1$, $\beta_{tot} = 1$, and $\beta_N = \frac{1}{2} + b$ with $b < 1/2$ (the exact values depend on the assumption on the area A; see Section 3.4). For the US, the values of these exponents are: $\beta_H \simeq 0.64$, $\beta_N \simeq 0.86$, $\beta_{tot} \simeq 1.03$, and $\beta_A \simeq 0.85$. As noted in Chapter 3, we observe several facts. First, the number of activity centers is described by a non-trivial exponent and this is the sign of a complex organization of the city, in between a monocentric structure and a completely decentralized case. Second, the area in any case increases sublinearly, in agreement with the general consensus about the density increasing with population (if $A \sim P^{\beta_A}$, then $\rho = P/A \sim P^{1-\beta_A}$). Third, the exponents for the total length of the network and the total commuting distance are different. It seems that the total commuting distance scales linearly with population, while the total length of the road network scales sublinearly.

Another theoretical approach was proposed by Bettencourt (2013), where the quantity of interest Y can be any "urban socioeconomic output" such as, for example, income. Bettencourt proposed a set of assumptions that are needed in this framework and we will discuss these:

- The economic output per capita Y/P is proportional to the average number of interactions $Y/P \sim g\langle k \rangle$. The quantity g is assumed to be constant and the average number of interactions is assumed to be $\langle k \rangle \sim \rho \sigma$, where σ is the "cross-section" of individuals.
- The cost of transport, assumed to be proportional to the total commuting length L_{tot}, is of the order of the minimum economic income: $Y_{min} \sim L_{tot}$. The quantity L_{tot} is given by ℓP, where ℓ is the average distance traveled.
- The average distance traveled ℓ depends on the fractal dimension H of the transportation network: $\ell \sim A^{H/d}$ (where usually $d = 2$ is the dimension of the embedding space).
- The total network length is given by the standard argument for spatial networks $L_N \sim \sqrt{AP}$.
- In order to compute the final exponent of the socio-economic income $Y \sim P^{\beta_Y}$, the area A is replaced by the length L_N.

These different assumptions lead to the following values for the main exponents

$$\beta_A = \frac{2}{2+H} \tag{7.98}$$

$$\beta_N = 1 - \delta \tag{7.99}$$

$$\beta_{tot} = 1 + \beta_A \tag{7.100}$$

$$\beta_Y = 1 + \delta \tag{7.101}$$

where $\delta = H/2(2+H)$ (in $d=2$ dimensions).

Indeed, if we assume that $Y_{min} \sim P^2/A \sim L_{tot} \sim \sqrt{\ell P}$, we obtain $A \sim P^{d/d+H}$. From there, we can estimate $L_N \sim \sqrt{AP} \sim P^{1-\delta}$ with $\delta = H/(2+H)$. At this point, we have $Y_{min} \sim L_{tot} \sim P^{1+2\delta}$. Bettencourt (2013) makes the additional replacement $A \to L_N$ and writes

$$Y \sim G \frac{P^2}{L_N}. \tag{7.102}$$

and obtains $Y \sim P^{1+\delta}$. The quantitative agreement is good but from a more theoretical point of view, taking transport cost proportional to the income seems to neglect other factors such as housing costs. Indeed, from the budget constraint for one household with income Y, transport and housing costs T and R reads

$$Y = z + T + R \tag{7.103}$$

where z is the composite commodity (see, for example, Section 7.1). It is clear from this constraint that the transport cost is not proportional to the income, but instead there is a variable share of housing expenditure depending on the income level.

In addition to these different problems, it seems rather difficult how to elaborate on this theoretical approach. Indeed, the discussions are essentially about scaling and this approach is not "mechanistic" in the sense where one could introduce various ingredients, write a general equation, and obtain some results. Even if this phenomenological approach is interesting and allows to discuss the effect of various forces acting in cities, it is difficult to use this model as a building block for further improvements.

8
Systems of cities

Regional science, followed by quantitative geography, described cities as systems integrating various layers and processes. In particular, the seminal paper of Christaller (1966) about the hierarchical organization of cities led to the "central place theory" that paved the way for a scientific approach to these problems, and constituted a pillar for thinking about the organization of cities. The influence of this paper was extremely important and we describe it here, with its limitations. In particular we discuss a result published by Okabe and Sadahiro (1996) which proposes a quantitative way to characterize a hierarchical system and shows that Christaller's results are actually true for a random system.

Probably the most striking quantitative fact about systems of cities is the Zipf law (Zipf 1949) which states that (see also Chapter 1)

$$P(r) \sim \frac{1}{r^\nu} \qquad (8.1)$$

where $P(r)$ is the population of the city at rank r (the largest city has rank $r = 1$) and ν is an exponent close to one. This result seems to be extremely robust and independent from the microdynamics that occur during the evolution of these systems of cities (Batty 2006). Such a robust, quantitative fact calls for a theoretical explanation and we discuss in this chapter the most important ones, including the Gabaix model (Gabaix 1999) and more recent approaches such as diffusion with noise.

8.1 Population distribution

For a given country, the population of cities follows a Zipf law characterized by an exponent that is generally close to one, and implies a number of simple consequences that we review here. In particular, we will discuss the population of the largest city, the ratio of the largest to the second-largest city, and the number of cities in a given country. Next, we will discuss some of the approaches for understanding this Zipf distribution. It is indeed a central problem for theoretical approaches to urban systems and has been tackled in many studies (see Batty 2006 and references therein). It is crucial to understand why we observe such a strong heterogeneity in population, while standard models in economics would predict an optimal size and therefore a peaked distribution of populations. There are many different models, but we will review the most important ones, starting with the Gibrat model and its variant proposed by Gabaix, and then focus on the diffusion model with noise, which seems to be a serious candidate for explaining not only the Zipf law but also the empirical diversity of the exponent value.

8.1.1 The number of cities and the largest city

The Zipf distribution has a certain number of consequences that have been discussed many times in the literature (see for example Pumain 2004 and references therein). In particular, we can easily estimate how the total number of cities varies with population and the size of the largest city in a country. The Zipf law for cities is equivalent to saying that the population distribution in cities is

$$\rho(P) \sim \frac{1}{P^{1+\mu}} \qquad (8.2)$$

(where $\mu = 1/\nu$; see Section 1.2). If we denote by N the number of cities and by U the total urban population in the country, we have the following relation:

$$S(N) = \sum_{i=1}^{N} P_i = U \qquad (8.3)$$

The exponent μ in Eq. (8.2) is around 1 and precise values are either just below or just above. When the exponent is $\mu > 1$, the dominant behavior of Eq. (8.3) is given by the usual central limit theorem

$$S(N) = U \sim N \qquad (8.4)$$

and the number of cities is thus proportional to U. When the exponent is smaller than one, we have a sum of broadly distributed variables with divergent average and the sum is dominated by its largest term and behaves (see for example Bouchaud and Georges 1990) as

$$S(N) \sim N^{1/\mu} \qquad (8.5)$$

8.1 Population distribution

which implies that the number of cities scales as $N \sim U^\mu$. We thus see that in both cases the number of cities varies as

$$N \sim U^\zeta \tag{8.6}$$

where ζ is either exactly equal to one or numerically close to it.

We now compute the distribution of the population of both the largest and second-largest city in a country. We sort the population of the N cities in decreasing order, $P_1 > P_2 > \cdots > P_N$, and we denote by $P_{\max} = P_1$ the maximum of this set. We also assume that these variables are independent and distributed according to the same law $\rho(P)$ with cumulative $F(x) = \text{Prob}(P < x)$. The cumulative distribution for the maximum is given by

$$P(P_{\max} = x) = NP(x)P(P_2 < x)P(P_3 < x)\cdots P(P_N < x) \tag{8.7}$$

$$\sim F(x)^{N-1} \tag{8.8}$$

$$\sim e^{-N\int_x^\infty \rho(y)dy} \tag{8.9}$$

where the last equality is valid for large N and large x. The average value of the maximum is such that the term in the exponential is of order one, which implies here that

$$\overline{P_{\max}} \sim N^{1/\mu}. \tag{8.10}$$

We see with this argument that what matters for the maximum is the tail of the distribution ρ (in fact, it can be shown that there are only 3 classes of extreme value distributions according to the decay behavior of ρ. For a modern treatment of this classical result, see for example Bouchaud and Mézard 1997).

More generally the rank of a value x is given by

$$r(x) = N\text{Prob}(P > x) = N \int_x^\infty \rho(x)dx \tag{8.11}$$

(which for the maximum correctly gives $r(P_1) \sim N \times 1/N = 1$). In the case of a broad law and for a large enough x, we thus recover the Zipf law for populations $r(P) \sim N/P^\mu$. The ratio of the largest city population P_1 to the second one P_2 is then given by

$$2 \sim \frac{N}{P_2^\mu} \frac{P_1^\mu}{N} \tag{8.12}$$

and we thus obtain for this "primacy index" (Pumain, 2004) the value $P_1/P_2 \sim 2^{1/\mu}$, which for μ close to one gives a ratio of order 2. This value can be compared, for example, to the case where variables are distributed according to a Gaussian law for which we would have $P_1/P_2 \simeq 1$. The fact that we have a broad law, Eq. (8.2), implies that the largest city is much larger than the second one, as would be expected for a peaked law.

Finally, we can easily estimate the number of cities (in a given country) with size larger than a value P and is then given by

$$N_>(P) = N \int_P^\infty \rho(x) dx \simeq \frac{N}{P^\mu}. \tag{8.13}$$

8.1.2 Gibrat, Gabaix, and diffusion with noise

There are many models that try to explain the Zipf law, such as self-organized criticality (Bak et al. 1987), highly optimized tolerance (Newman 2000), random killing time of stochastic processes (Reed and Hughes 2002), or spatial analog of combinatorial processes (Aldous and Huang 2012), but there is no clear answer to that problem yet. In this section, we discuss models that can explain this power law and that are based on reasonable processes from the point of view of cities. Indeed, in order to be a good model it is not enough to reproduce the power law, but the mechanisms invoked have to be realistic and likely to occur in reality. This is one of the reasons why we present only a few models here. First, we discuss the Gibrat model and its extension proposed by Gabaix (1999), as it is the model that is the most commonly accepted. Second, we discuss a different approach that contains the physics of the Gibrat model but also makes the connection between this problem and a class of diffusion problems (diffusion with noise, or stochastic diffusion) widely studied in statistical physics.

The Gibrat and Gabaix models

Gibrat (1931) proposed a simple rule stating that the growth rate of a firm is independent of its size. Applied to the growth of city populations, it leads to the following equation

$$P_i(t+1) = \eta_i(t) P_i(t) \tag{8.14}$$

where $P_i(t)$ is the population of a city i at time t and where the growth rate η_i is random but independent of the size P_i. In the following we normalize all quantities by the total population of all cities and consider the quantity

$$w_i(t) = \frac{P_i(t)}{\overline{P}(t)} \tag{8.15}$$

where the denominator represents the average population at time t. Here all cities are independent from each other and we drop the index i in all quantities. If we iterate Eq. (8.14), we obtain $P(t) = P(0) \prod_0^t \eta(t)$ and by taking the logarithm, we obtain

$$\log w(t) = \sum_{\tau=0}^{t} \log \eta(\tau) + \log P(0) \qquad (8.16)$$

(we assume that $P(0)$ is small and neglect it in the following and that $\overline{P(t+1)/P(t)}$ is a constant rescaling η). The sum of random variables follows the central limit theorem (for reasonable choices of the variables η) and the distribution of w is then a lognormal distribution of the form

$$\rho(w) = \frac{1}{\sqrt{2\pi\sigma^2 t}w} e^{-(\log w - vt)^2/2\sigma^2 t} \qquad (8.17)$$

where $v = \langle \log \eta \rangle$ and $\sigma^2 = \langle (\log \eta)^2 \rangle - \langle \log \eta \rangle^2$ (the brackets denote the average over the distribution of the random variable η). Numerically, this distribution can be confused with a power law (with exponent 1). Indeed, this distribution can be rewritten (Sornette and Cont 1997) as

$$\rho(w) = \frac{1}{\sqrt{2\pi\sigma^2 t}} \frac{1}{w^{1+\mu(w)}} e^{\mu(w)vt} \qquad (8.18)$$

where the "effective exponent" reads

$$\mu(w) = \frac{1}{2\sigma^2 t} \log \frac{w}{e^{vt}}. \qquad (8.19)$$

Since the function μ is slowly varying with w we could see (numerically) an apparent power law with exponent $1 + \mu$, but this ambiguity disappears for large w. In addition to this numerical problem, there is the difficulty (see Chapter 1) of defining cities which have an impact on the tail of the distribution. Despite these various problems, recent papers (Soo 2005; Rozenfeld et al. 2008; Malevergne et al. 2011), building up on different city definitions (such as the clustering methods discussed in Section 1.1) show convincingly that the population distribution for cities is indeed in the power-law regime. In particular, it was shown by Rozenfeld et al. (2008) that the growth of urban areas is not consistent with Gibrat's law, and that the mean and standard deviation of the growth rate depends on the city size.

In order to understand the Zipf law, Gabaix (1999) proposed a variant of the Gibrat model based on the idea that small cities cannot shrink to zero. In other words, the process considered by Gabaix is a random walk with a lower reflecting barrier which can, in some conditions, produce a power-law distribution of populations. There are two essential assumptions here. The first ingredient is the existence of a perfectly reflecting barrier, which prevents cities from becoming too small. The second ingredient is that $\langle \log \eta \rangle < 0$. Indeed, if that is not the case, the walk will escape to infinity and we recover the central limit theorem and a lognormal distribution for w. Instead, if the walker is pushed towards zero ($\langle \log \eta \rangle < 0$), it will bounce on the reflecting barrier and create an effective flow going to large values. This interplay between a drift towards the barrier and the reflection at the barrier is what gives rise to a power law in this model.

This problem of a multiplicative noise with a reflecting barrier was solved exactly by Levy and Solomon (1996) and discussed in more detail by Sornette and Cont (1997). We change variables $x = \ln w$ and $\ell = \ln \eta$ and we assume that the reflecting wall is at $x_0 = \ln w_0$. We propose here a simple derivation for computing the distribution of x [1]. In this process, the position of the walker at time $t+1$ is given by (we assume that time is discretized)

$$x_{t+1} = \max(x_0, x_t + \ell) \tag{8.20}$$

where ℓ is the value of the increment (and distributed according to $\pi(\ell)$). This equation means that $x_{t+1} = x_0$ if the jump brings the walker below the barrier, and $x_{t+1} = x_t + \ell$ is the new position if it is above the barrier. The walker will then be at x_0 with probability

$$P(x = x_0) = \int_{-\infty}^{x_0} dx \int_{-\infty}^{+\infty} P(x - \ell) \pi(\ell) d\ell \tag{8.21}$$

where $P(x - \ell)$ is the probability of being at position $x - \ell$ and $\pi(\ell)$ is the probability of jumping a distance ℓ. For $x > x_0$, the probability is then

$$P(x) = \int_{-\infty}^{+\infty} P(x - \ell) \pi(\ell) d\ell. \tag{8.22}$$

This last result is a Wiener–Hopf equation and solving it can be mathematically involved. Here, we proceed in a simplified way and assume an exponential ansatz $e^{-\mu x}$ for the solution. Plugging this form into Eq. (8.22) leads to the equation that μ must satisfy:

$$1 = \int e^{\mu \ell} \pi(\ell) d\ell \tag{8.23}$$

The stationary limit is then given by $P(x) = e^{-\mu x}$, which implies that the distribution of w is given by a power law of the form

$$P(w) \sim \frac{1}{w^{1+\mu}}. \tag{8.24}$$

---

[1] I thank K. Mallick for giving me this argument.

This result was discussed by Sornette and Cont (1997), who proposed another derivation for it. In particular, they proposed the following intuitive explanation for the appearance of a power law: in the presence of a reflecting barrier which prevents the random walk from escaping toward very small values, there is a continuous flow of particles that can sample the large positive values of x, leading to a broad tail for P.

The exponent μ satisfies Eq. (8.23) and in the limit where the distribution $\pi(\ell)$ is not far from a Gaussian, we have the following approximate expression (Sornette and Cont 1997):

$$\mu \simeq |v|/\sigma^2 = |\langle \log \eta \rangle|/\langle (\log \eta)^2 \rangle - \langle \log \eta \rangle^2. \tag{8.25}$$

At this point, even if the final result is satisfying, it seems difficult to explain the reason for the existence of a minimal value of the population P (or w) and even so, how we could justify that $\langle \log \eta \rangle < 0$. In the next section we discuss another mechanism that can lead to a power law and that seems more realistic for cities.

Diffusion with noise

The "regularization" of the Gibrat model proposed by Gabaix relies on the assumption of a minimum size for cities. However, we know cases of shrinking cities, and even cities that disappeared completely. A different, simple way to "regularize" this behavior is to introduce migration effects between cities and we will follow the presentation given by Bouchaud and Mézard (2000). The equation for the population $P_i(t)$ of city i at time t is given by

$$\frac{dP_i}{dt} = \eta_i(t) P_i(t) + \sum_j J_{ji} P_j(t) - J_{ij} P_i(t) \tag{8.26}$$

where the first term represents the "internal" growth (corresponding to the Gibrat model) and where the last two terms represent movements between cities (see Fig. 8.1). The random variables $\eta_i(t)$ are assumed to be identically independent Gaussian variables with the same mean m and variance given by $2\sigma^2$ (we kept the

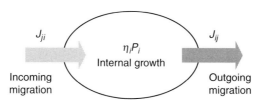

Figure 8.1. Schematic representation of Eq. (8.26). There is an internal flow and "migrations" going either out or in are described by the matrix J_{ij}.

same notation as in Eq. (8.14) even though there is a difference of one). The flow (per unit time) between cities i and j is denoted by J_{ij} and for a general form of these couplings we are unable to solve this equation. We can, however, discuss the simple case of the complete graph where all units are exchanging with all others at the same rate, taken equal to $J_{ij} = J/N$, where N is the number of cities (this scaling ensures a well-behaved large N limit). The equation for P_i becomes

$$\frac{dP_i}{dt} = J(\overline{P} - P_i) + \eta_i(t) P_i \qquad (8.27)$$

where $\overline{P}(t) = \sum_i P_i(t)/N$. We see that the first term acts as a homogenizing force towards the average $\overline{P}$. All cities feel the same environment, which shows the mean-field nature of this simplified case. Formally we can treat this equation as an equation for P_i subjected to a source $\overline{P}(t)$, and by integrating formally we obtain

$$P_i(t) = P_i(0) e^{\int_0^t (\eta_i(\tau) - J) d\tau} + J \int_0^t d\tau \overline{P}(\tau) e^{\int_\tau^t (\eta_i(\tau') - J) d\tau'}. \qquad (8.28)$$

The average quantity $\overline{P}$ is still unknown at this point and if we sum this equation over i we still cannot solve it. However, if we assume that in the large N limit, the quantity $\overline{P}$ is self-averaging

$$\overline{P} \approx_{N \to \infty} \langle \overline{P} \rangle \qquad (8.29)$$

where the brackets $\langle \cdot \rangle$ denote the average over the variable η, we have

$$\overline{P(t)} \simeq \overline{P(0)} e^{(m+\sigma^2 - J)t} + J \int_0^t d\tau \overline{P(\tau)} e^{(m+\sigma^2 - J)(t-\tau)} \qquad (8.30)$$

(we assumed here that the initial conditions are independent of η). This equation can be easily solved (by Laplace transform, for example) and we obtain

$$\overline{P(t)} = \overline{P(0)} e^{(m+\sigma^2)t}. \qquad (8.31)$$

In order to observe a stationary distribution we normalize P_i by $\overline{P(t)}$ and construct as above the variables $w_i = P_i(t)/\overline{P(t)}$. Using Eq. (8.27) these quantities obey the following Langevin-type equation

$$\frac{dw_i}{dt} = f(w_i) + g(w_i) \delta \eta_i(t) \qquad (8.32)$$

where $f(w) = J(1-w) - \sigma^2 w$ and $g(w) = w$ (and $\delta \eta = \eta - m$). In order to write the Fokker–Planck equation for the distribution $\rho(w)$, we have to give a prescription about the multiplicative noise. For this we must specify the correlations in the product and there are basically two main prescriptions, the Ito and Stratonovich ones. We choose here the Stratonovich prescription, which usually corresponds to the most physical one, and obtain the Fokker–Planck equation for the distribution $\rho(w)$ under the form

$$\frac{\partial \rho}{\partial t} = -\frac{\partial}{\partial w}[f\rho] + \sigma^2 \frac{\partial}{\partial w}\left[g\frac{\partial}{\partial w} g\rho\right]. \tag{8.33}$$

The equilibrium distribution which satisfies $\partial \rho/\partial t = 0$ obeys a simple equation, and once solved leads to

$$\rho(w) = \frac{1}{\mathcal{N}} \frac{e^{-\frac{\mu-1}{w}}}{w^{1+\mu}} \tag{8.34}$$

where $\mathcal{N}$ is a normalization constant and where the exponent is given by

$$\mu = 1 + \frac{J}{\sigma^2}. \tag{8.35}$$

This result implies a number of consequences. First, we observe that this regularization changes the lognormal distribution to a power law (at large w) with exponent between 1 and 2 (for J/σ^2 small). This distribution has a finite average $\langle w \rangle = 1$, but with an infinite variance, signalling large fluctuations. This regularization thus provides a simple explanation for the diversity of exponents observed in various countries (see Fig. 1.10 and Soo 2005). The result is similar to what is obtained with the Gabaix model, but here we obtain an exponent whose value depends on migration between cities. The diffusion model discussed here shows that the origin of the Zipf law could lie in the interplay between internal random growth and exchanges between different cities. An important consequence is that increasing mobility should actually increase μ and therefore reduce the heterogeneity of the city size distribution. A few words of caution are however needed here: first the relation Eq. (8.35) should be tested empirically, and second, the important assumption that J_{ij} is constant is by no means obvious and we could imagine that this quantity actually depends on the distance $d(i, j)$ which could alter the results.

Correlations in the demography dynamics

In addition to the different problems discussed above, the Gibrat model actually also neglects correlations between cities, an assumption that can now be tested. The study of spatial correlations is not new; as early as the 1970s Glass and Tobler (1971) considered the spatial distribution of cities in the Spanish plateau around Madrid. They computed the radial distribution $g(r)$, defined as the probability that a city will be at a distance r from another one, and observed a repulsion at short distances and damped oscillations for larger distances, consistent with the image of hard-core spheres.

More recently, Hernando et al. (2014, 2015) focused on spatial and temporal correlations of populations for Spain and the US. For the the US these authors use a large database that provides 170 years of data (1830–2000) for the evolution of counties. Even if counties can contain more than one important urban center, this study can however bring some useful information about spatio-temporal correlations of urban population growth given by

$$c(t, t+\Delta t) = \langle \dot{P}_i(t) \dot{P}_j(t+\Delta t) \rangle - \langle \dot{P}_i(t) \rangle \langle \dot{P}_j(t+\Delta t) \rangle \tag{8.36}$$

where the population growth rate is $\dot{P}_i(t) = dP_i/dt$ ($P_i(t)$ is the population of county i at time t). The brackets denote here the average over all counties $\langle O \rangle = \sum_{i=1}^{N_c} O_i/N_c$ (where N_c is the number of counties). This correlation, averaged over all possible initial times

$$\langle c(\Delta t) \rangle = \frac{1}{N(t)} \sum_t c(t, t+\Delta t) \tag{8.37}$$

(where $N(t) = 12$ is the number of different times available), is shown in Fig. 8.2(a) and exhibits a typical correlation time τ of the order of a few decades (an exponential fit gives $\tau \approx 25$ years).

The spatial correlations of population growth can be characterized in the same way by spatial correlation coefficients. For this, we first estimate the correlation between growth rates of counties i and j

$$C(i, j) = \overline{\dot{P}_i(t) \dot{P}_j(t)} - \overline{\dot{P}_i(t)}\, \overline{\dot{P}_j(t)} \tag{8.38}$$

where the bar $\overline{\cdot}$ is the average over time. We then average over all counties at distance d

$$C(d) = \frac{\sum_{i,j} C(i, j) \delta(d(i, j) - d)}{\sum_{i,j} \delta(d(i, j) - d)} \tag{8.39}$$

($d(i, j)$ is the Euclidean distance between counties i and j). This quantity is represented in Fig. 8.2(b) and displays a slow decay with a typical correlation distance of order $d_0 \approx 200$ km for the US, while $d_0 \approx 80$ km for Spanish cities (not shown). We note that these correlation lengths are actually approximately in the same ratio as the typical size of these countries. Indeed $d_0(\text{US})/d_0(\text{Spain}) = 200/80 \simeq 2.5$ is of the same order of magnitude as $\sqrt{A_{US}/A_{Spain}} \approx 4.5$ (where A represents the area of the country) which seems to indicate that the evolution in the US is not more coherent than in Spain but is adapted to the system size.

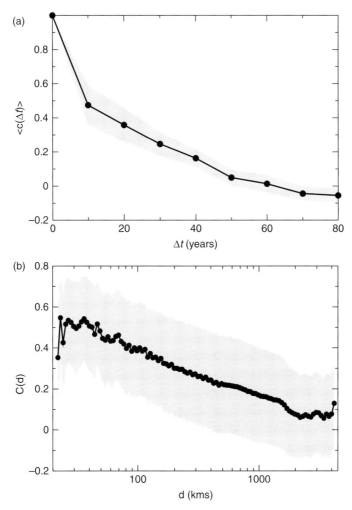

Figure 8.2. (a) Average growth rate correlation versus time for counties in the US with more than 10,000 inhabitants (the shaded area represents one standard deviation limit); (b) spatial correlation $C(d)$ for pairs of US counties at distance d. The dark line is the result obtained for US counties (and the shaded areas represent the one standard deviation limit). More details can be found in Hernando et al. (2015). Figure and data courtesy of A. Hernando.

These indications show that in order to understand the evolution of a city, we need to mention temporal and spatial resolution. In general, we probably cannot assume that cities are isolated but embedded in a system of cities, as is well understood and discussed by geographers (see, for example, Rozenblat and Pumain (2007) and earlier references therein). At a relatively small temporal or spatial scale, the population of cities seems to be strongly correlated, a correlation that disappears for larger distances.

Finally, we mention that a further test in order to distinguish various models could be to compare with data their predictions for these correlations. Time correlations between different cities are absent in the Gibrat model, but the diffusion model Eq. (8.27) predicts time correlation functions that can be computed (Bouchaud and Mézard 2000) and compared with data. Further studies are obviously needed and hopefully the future will tell us which model provides the correct explanation for the Zipf law.

8.2 Central place theory and spatial fluctuations

The idea that there is some underlying organization in a system of cities was proposed by Christaller (1966) in his famous "Central Place theory" (the German edition of the paper appeared in 1933). As mentioned by Krugman (1996), the central place theory is not really an (economic) model as it does not provide an explanation in terms of linking microscopic motives and macroscopic emergent behavior. This, however, did not prevent many authors and scientists working on cities and systems of cities to think cities with the framework of this central place theory. For this reason, it is important to discuss it in detail and to highlight its major drawbacks.

We will also discuss a statistical approach to the idea of hierarchical organization of places, based on the paper by Okabe and Sadahiro (1996) who proposed interesting tools in order to characterize such a structure. They also showed that Christaller's observations could actually result from randomness and not from organizational principles. This discussion highlights the importance of a null model in order to understand fluctuations and their relevance.

8.2.1 Outline of Christaller's theory

The main purpose of the central place theory is to explain the size, the number, and the spatial organization of human settlements. The settlements are considered as central places that provide services to the surroundings. There is a large number of assumptions considered by Christaller such as the homogeneity of space, or a uniform population density. Probably the most important assumption is that every customer should be in a zone served by only one market area, leading to triangular or hexagonal arrangements of settlements. The interesting idea in the central place theory is to connect the spatial location with another attribute such as population in the case of the hierarchical organization of towns and cities. We can indeed expect a hierarchy of places which are not organized at random: a very large center is surrounded by medium-sized centers which themselves are surrounded by small centers, and so forth. There is then a clear relation between the population ranking and the spatial arrangement of units such as towns, activity centers, etc.

More precisely, following Okabe and Sadahiro (1996), there are three main results from Christaller's theory:

- C1. The shape of each dominant region is a regular hexagon.
- C2. The dominant regions at the same level have the same area.
- C3. There is a hierarchy of centers and at each level i of the hierarchy there are n_i corresponding local centers. The ratio

$$K_i = \frac{n_i}{n_{i-1}} \tag{8.40}$$

is constant, $K_i = K$.

This value of K is the fundamental parameter in Christaller's approach and it was proposed that different values of K correspond to different hierarchical organizations: for $K = 3$ we have the hierarchy of the "marketing principle," for $K = 4$, the "transport principle," and for $K = 7$ the hierarchy of the "administrative principle" (Christaller 1966). We can understand the value of K in terms of the ratio of market area between nodes that are at different levels of the hierarchy. For example, in the marketing principle $K = 3$, the nodes of different levels are located on the hexagonal lattice in such a way that the market area of a higher-level node occupies an area A_0 which is three times smaller than one of lower-level node (and similar considerations for $K = 4$, $K = 7$).

8.2.2 Spatial fluctuations

The impression and the idea of central place theory were so strong and appealing that the paper by Okabe and Sadahiro which appeared in 1996 remained largely overlooked. In this paper, "An illusion of spatial hierarchy: spatial hierarchy in a random configuration," Okabe and Sadahiro are studying fluctuations in random spatial distribution of points and show that what Christaller thought of as patterns could in fact result from randomness.

Spatial distributions and their characterization such as tesselations are so important that we reproduce and discuss the arguments of Okabe and Sadahiro here. In addition, we believe that the tools developed in this paper – in particular for characterizing a spatial hierarchy – can be useful to scientists working on spatial systems. Okabe and Sadahiro consider the general problem of points randomly distributed in 2d space. These points represent activity places and can be towns or regions. We start from N points that are distributed in the plane with coordinate x_i ($i = 1, \cdots, N$) and we assume that they are also described by an attribute a_i,

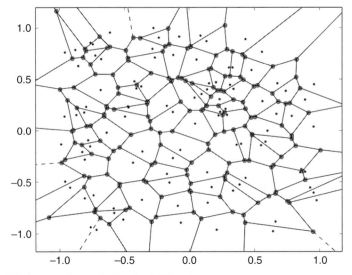

Figure 8.3. Voronoi tesselation for 100 nodes.

which can be the population or any other quantity that allows one to rank these nodes.

The main idea is to first construct a Voronoi tesselation around the points. The Voronoi cell V_i of a node i is the set of points that are closer to i than to any other node:

$$V_i = \{\, x \mid d(x, x_i) < d(x, x_j), \forall j \neq i \,\}. \tag{8.41}$$

We show in Fig. 8.3 an example of a Voronoi tesselation computed for 100 nodes distributed uniformly in the plane. In addition here, the nodes can also be sorted according to their attribute but the main difficulty – as pointed out by Okabe and Sadahiro (1996) – is to characterize mathematically the notion of spatial dominance and of a spatial hierarchy. They proceeded in three steps:

1. Find the local centers.
2. Rank the local centers.
3. Determine the spatial relations among local centers.

For the first task, they define local centers according to the very intuitive idea of local maxima: for any point i we can define, with the help of the Voronoi tesselation, the set of neighbors $\Gamma(i)$ that are the nodes whose Voronoi cell is adjacent to V_i. A node i is then a local center when its attribute a_i is larger than the attributes of its neighbors

$$i \text{ is a local center} \iff a_i > a_j, \ \forall j \in \Gamma(i). \tag{8.42}$$

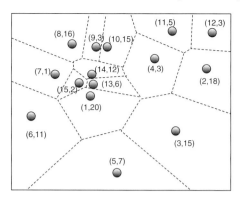 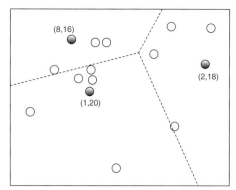

Figure 8.4. Different levels of local centers. In the left figure, nodes are represented by (i, a_i) where i is the index of the node, and a_i the corresponding attribute (such as the population, for example). The dotted lines represent the Voronoi tesselation constructed on the nodes present at that level. In the right panel we represent the local centers at the first level and the corresponding Voronoi tesselation.

Starting from an initial configuration of nodes, denoted by $P^{(0)}$, we can then construct the set of local centers $P^{(1)}$, and continue this process recursively until there is only one node left at a certain level m (see a simple illustration of one step of this process in Fig. 8.4). We thus have a series of sets which satisfy $P^{(m)} \subset P^{(m-1)} \subset \cdots \subset P^{(1)} \subset P^{(0)}$. The number of local centers at each level naturally decreases and if the number of neighbors is roughly constant $|\Gamma|$, we expect an exponential decrease

$$|P^{(k)}| \sim N e^{-k \log |\Gamma|} \tag{8.43}$$

which implies that for N initial nodes, there are about $m \sim \log N$ levels in order to reach one single local center. We can now define the rank of local centers. The first-rank center is the node left at the last m-th level. The second-rank centers are those that are present at level $m-1$ but not at level m. In general, the centers of rank j are present until level $m-j+1$, but not at level $m-j+2$.

We now have all the tools for defining the spatial dominance according to Okabe and Sadahiro (1996). The starting idea is that in the Voronoi tesselation all points inside a given Voronoi cell i are "dominated" by this node i, in the sense that it is the nearest point in $P^{(0)}$ to all these points. From a marketing point of view, the consumers in V_i minimize their distance to the facility located in i. Following this idea, a local center of level $k-1$ is spatially dominated by the local center i if it is included in the Voronoi cell of i at level k. The spatial dominance allows to construct a tree for all the nodes (see Fig. 8.5) which represents the various levels of centers and their relations. We note that since it is a tree (and not a complete graph), there is a spatial dominance relation only between certain nodes but not among all

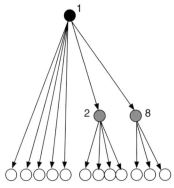

Figure 8.5. Tree graph that represents the spatial hierarchy of nodes in the example of Fig. 8.4.

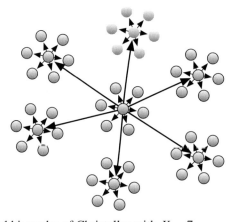

Figure 8.6. Spatial hierarchy of Christaller with $K = 7$.

of them. In particular, for the Christaller $K = 7$ case, the spatial dominance relation can simply be represented as in Fig. 8.6.

We can now characterize the spatial hierarchies for any distribution of points. In particular, it is important to do this analysis for the uniform distribution of points where the locations x_i are randomly distributed in the plane (in a bounded region) and attributes are integers attributed randomly to each node. We will now discuss the three main results of Christaller (C1–3) for this random null model.

First, the general properties of planar Poisson–Voronoi tesselations are by now well known (see Hilhorst 2008 and references therein). In particular, the probability distribution $p(n)$ to have a Voronoi cell with n sides has been discussed in many studies. This quantity has a peak at $n = 6$ and decreases quickly with n as

Hilhorst (2005)

$$p_n \sim \frac{C}{4\pi^2} \frac{(8\pi^2)^n}{(2n)!} \left[1 + \mathcal{O}(\frac{1}{\sqrt{n}})\right] \qquad (8.44)$$

where C can be computed exactly and is $C = 0.344\cdots$. The average value of n is $\langle n \rangle = 6$, which corresponds indeed to the intuitive idea that most Voronoi cells are hexagons, but there are obviously other polygons distributed according to $p(n)$. This well-known result is consistent with the first (C1) observation by Christaller.

The second result in Christaller's theory (C2) concerns the area of polygons. In the decimation process when we go from the initial state to a higher level, we note that local centers at level k cannot be neighbors at level $k-1$. In other words there is a buffer zone between them and their distribution is therefore not a simple Poisson process. In fact this effective repulsion effect tends to homogenize their distribution and leads to more homogeneous Voronoi cells. This idea was tested against Monte-Carlo simulations in Okabe and Sadahiro (1996) and although further tests are probably necessary, they observe a reduction of the relative dispersion of area during the decimation. At the level 0, they computed a relative dispersion of 0.53, which is in agreement with their numerical calculations, and at the first level they found a value of order 0.35. Here also, the result (C2) of Christaller is actually not inconsistent with a random distribution.

Okabe and Sadihiro also tested the last result (C3), which states that the ratio of the number of local centers of two consecutive levels is constant. Okabe and Sadahiro (1996) found results in agreement with Christaller's theory (for the "administrative principle") with a value around $K \approx 6.7$, although further simulations are needed in order to make this statement more precise.

At this point the most important quantitative statements in the central place theory actually appear to be consistent with a random (Poisson) Voronoi tesselation and this has important implications. First, even if Christaller's results seem to be correct, it is not because of some underlying mechanisms that control the organization of these systems. Rather, randomness and spatial constraints seem to be the most important features here and even if further simulations are certainly needed, the Christaller "theory" is at this point not clearly distinguishable from a purely random process. This might also be the reason why these spatial hierarchies often seem to appear in the real world. From a methodological point of view, this example shows the importance of null models when trying to estimate the importance of an effect.

9
Toward a new science of cities

9.1 What is our "understanding"?

As we discussed in Chapter 1, "understanding" has many definitions, with a variable amount of quantitative input. For a physicist, understanding does not mean having a story consistent with reality only, but also having mathematical tools and models able to describe real phenomena and to predict the outcome of experiments. Even if a qualitative description of processes is somewhat satisfying, it is not enough for constructing a science of cities. Indeed, we would like to identify the most important parameters, not only to understand the past, but also to be able to construct a model that gives, with a reasonable confidence, the future evolution of a city and to test the impact of various policies.

At this point, we certainly have a number of pieces of the puzzle, and we have discussed some of them in this book. It doesn't however mean that we have solved the full puzzle. New data sources and large datasets allow us to get a precise idea of what is happening in cities. We are currently experiencing an exciting time during which we can challenge the purely theoretical developments made these last decades. In many empirical studies, the identification of relevant factors was essentially done statistically, and we can now hope to go beyond and to have a more mechanistic approach, where a model based on simple processes is able to reproduce empirical observations.

Concerning the spatial structure of cities, new data sources give us a real-time, high-resolution picture of mobility. The structure of mobility flows that come out from these datasets departs from the usual image of a monocentric city where flows converge towards the central business district. Instead, for large cities, the main flows appear to be far from the localization between centers of residence and activities, that we could have naïvely expected. This massive amount of data also allows us to quantitatively assess the degree of polycentricity of an urban system. A simple model showed that congestion is a crucial factor

9.1 What is our "understanding"?

in understanding the evolution of polycentricity and mobility patterns with the population size. In order to go deeper in our understanding of the spatial structure of cities, infrastructures such as streets and subways play an essential role. Their durability affects the city's landscape and organization, and understanding the evolution of these networks is another crucial ingredient for a science of cities. We discussed how to characterize these structures and described some directions for modeling their evolution. In particular, we explored how a cost-benefit approach is helping to explain the properties of subway networks and their relation to socio-economical indicators.

In addition, mobility is not unimodal anymore and an increasing share of trips in large cities is made with two or more transportation modes. Multimodality affects the patterns of mobility and also poses new questions about the coupling between these different modes. We showed that the multilayer network approach proposes a convenient framework for characterizing these systems and their efficiency.

Mobility is related to another major dimension in cities, which is the economic activity and the spatial density of jobs. Surprisingly enough, the first empirical results reveal some similarities in the commuting behavior of individuals in the United States and the UK (and Denmark within a certain variation range), despite many differences in the organization and structure of these countries. This has to do with the decision process of individuals with respect to job and residence choices. There is obviously a part for rationality, but many other factors play an important role and the modeling task is very complex. Surprisingly enough, an approach based on a simple stochastic process explains the appearance of this almost "universal" commuting behavior. Instead of trying to describe precisely all processes, it might be a good idea to try out simple models where complex quantities are replaced by random variables and complex processes by stochastic models. This approach proved to be useful in the physics of a large number of interacting particles and might find interesting applications in social and economic systems.

Obviously a city is not about mobility and transport infrastructure alone, and a central aspect is how the social structure affects its organization and how economic activities are distributed in space. These are rather formidable problems, but we have tried to show in this book that there is still some need to define the proper tools for measuring properties, even for some very well-studied phenomenon such as segregation. Simple models with a deep connection with statistical physics such as the Schelling model also lead to new ideas and demonstrate the failure of long-lived beliefs such as Adam Smith's invisible hand.

Concerning systems of cities, the origin of the observed Zipf law for populations could actually reside in exchanges between cities. At this point, more data about inter-urban migration are necessary in order to test thoroughly this idea, but it provides a simple guide for elaborating further theories about interacting cities. In addition to Zipf's law, the existence of nonlinear scalings with non-trivial exponents can serve as a guide toward the identification of dominant mechanisms governing the evolution of urban systems. However at the time this book is written it seems that the robustness of many of these exponents is rather weak. Many of their values are very close to 1 and in some cases the noise in the data leads to low confidence in these values. Nonlinear scaling is certainly a very important and useful tool for constructing a scientific approach to cities, but the empirical determination of exponents should be very clean and non-ambiguous. If it is not, any theoretical construct might fall apart like a house of cards.

9.2 Measuring the death and life of great cities

Despite all the efforts made by many different scientific communities, it is fair to say that our understanding of cities is still limited. The availability of massive amounts of data about all sorts of processes might give us a chance to change that. The most difficult challenge is certainly to integrate all the different approaches and to extract the relevant ingredients for all these processes. Extracting the main ingredients, the dominant mechanisms give some kind of blueprint upon which more detailed simulations can be built.

As already emphasized in many parts of this book, data will change our way to study cities and in particular will help us for testing old ideas about them. It thus seems relevant to end this book with a note on what is considered a pillar of city planning. In her famous book *The Death and Life of Great American Cities*, Jacobs (1961) strongly criticizes American urban planning policy in the post-war era and exhibited four conditions that have to be simultaneously satistified for promoting life in cities:

- high density
- mixed land use, with different activity times
- small blocks
- diversity in buildings.

Even if these conditions seem perfectly reasonable, Jane Jacobs essentially reached these conclusions from subjective grounds. The exciting perspective with new sources of data gives the possibility of actually quantitatively testing these assumptions (Sung et al. 2015; Nadai et al. 2016). In particular, Nadai et al. (2016) used cell phone data for cities in Italy together with other data sources (open street map, census, land use, infrastructure, and Foursquare data), in order to thoroughly test these assumptions. They were using cell-phone activity as an indicator for urban vitality (defined as the number of internet connections per day and per unit area) and other data (such as Foursquare) to determine the urban diversity. The variables used to test Jacobs' assumptions can be classified into two categories. The first category characterizes land, building use and concentration. The second category of indicators concerns the activity at each location, and measures whether there is some nightlife or commercial activity, and whether there are places that foster communication. Very interestingly, the main result of this study is that indeed vitality and diversity are strongly connected and that the four conditions enumerated by Jacobs are definitely very important.

These datasets exist for many cities in different parts of the world and open the door to testing quantitatively a number of indicators, paving the way to a quantitative urbanism, strongly grounded in data. Even if the theory is not at this level of pragmatism, this example of urban computing (Zheng et al. 2014) actually adds an optimistic note about the possibility of building a science of cities.

9.3 The future of the city

Many scenarios for the future of cities have been proposed by various think-tanks or companies, but even if they sometimes look reasonable their scientific validity is questionable. Most of these scenarios depend on the taste of their authors and on what they think is relevant. As a result, most of their projections depend largely on ideologies and opinions and not on a quantitative understanding of cities.

An important problem is of course to mitigate bad effects and to understand the consequences of urban projects. This amounts to an understanding of the main forces at play and the interplay between natural tendencies and governance. So far the data are relatively clear: there is an important increase of bad indicators such as the CO_2 emitted by transport or the total time spent in traffic jams. All things being equal, if we extrapolate the current curves we find a point in the near future where the amount of delays cannot be sustained. Electric cars could solve the small particles and CO_2 emission problem (although we have to include the whole chain of car production, including the pollution generated by batteries), but certainly not congestion problems and the time spent in traffic jams. This is a major problem, as flows in cities are fundamental for their good functioning. Congestion is thus a good indicator of the economic health of a city, and when individuals cannot find a job close (in terms of trip duration) to their residence, or a home close to their office, it is the sign of many problems.

Polycentricity seems to be one of the "natural" reactions of cities to congestion and we can estimate for the United States what would happen if cities had stayed monocentric. The ratio of delays spent in traffic jams for the polycentric city and the corresponding monocentric case (if the city had stayed monocentric but with the same population) is given by (Louf and Barthelemy 2014a)

$$\frac{\delta \tau_{\text{mono}}}{\delta \tau_{\text{poly}}} \sim \left(\frac{P}{c}\right)^\zeta \qquad (9.1)$$

where the exponent is of order $\zeta \approx 0.57$. Even if the total delay grows superlinearly in the polycentric case, it would then have been much worse in the monocentric case with a growth (for the US) of order $P^{1.9}$! Polycentricity then naturally reduces the total delay and consequently bad side effects such as particle and CO_2 emissions.

Nonetheless, if the individual car stays the dominant transportation mode, cities will put more strain on people's lives, while acting as a catalysts for the production of CO_2 greenhouse gas and small particles responsible for health problems. It is currently believed that the advantages associated with living in a large city outweigh the costs. Scaling results reveal however the existence of very rapidly growing problems such as congestion and CO_2 emissions, which inevitably begs the question of the sustainability of large cities. It might be time to cut down considerably the use of individual vehicles, or to consider the possibility of living in smaller or medium-sized cities: the infrastructure costs may be larger, but the impact on the environment and on the well-being of people would be globally beneficial. From the point of view of mobility and congestion, the notion of the urban village seems then to be the most sustainable. A plan that is too rigid is very likely not well adapted to the evolution of cities. We saw examples of cities essentially planned for cars – such as Brasilia, for example – that could not sustain on large increase in population, leading to a chaotic development of suburbs. In this respect, it is surprising that some current projects such as the King Abdullah Economic City (Saudi Arabia) are based on very strict spatial delimitations of land-use. This organization will probably work as long as the population does not increase too much, but there is a real danger of mid- to long-term failure, if the city develops itself too well!

The digital revolution can of course not be overlooked and all its consequences on the structure of societies and cities are still difficult to foresee entirely. In particular, an important game changer will probably be remote working. This seems to be a growing trend and has the potential to trigger crucial changes in the organization of work, and to alter the shape of cities. Cities existed in the first place for spatial reasons and if we remove this distance component from the equation, we can expect crucial changes. Space will, however, not be removed completely: products will still need to be assembled and transport costs won't disappear. The importance of space depends a lot on the nature of the activity and the digital revolution could reinforce spatial segregation and heterogeneities in our societies.

9.4 Concluding thoughts

We now have data about almost every facet of urban systems, and we can try to construct a theory in agreement with these data. However, a lot of work is needed in order to integrate information and knowledge about all these aspects and the network of cities on a larger scale. This huge complexity suggests that the modeling of cities is certainly limited and that we cannot expect to obtain a closed theory of urban systems, able to discuss simultaneously all phenomena taking place in cities. We can expect to be able to determine the main parameters and ingredients for a given question, without forgetting that many aspects are correlated. For example, in physics we are able to discuss the conductivity of materials without discussing their mechanical properties, even if the coupling between these two aspects leads to new effects such as piezoelectricity. The difficulty is then evident for cities. We have to carefully choose the ingredients, making sure that we understand couplings with other processes. This ambitious goal is the key to our quantitative understanding of cities and the possibility of modeling them.

The goal of a science of cities would be to understand how cities grow and eventually to provide scientific support for urban planning operations. As we have already discussed, there are so many processes at play in urban systems that this program is a rather formidable task. This book advocates the use of simple models, guided by ideas of the statistical physics of complex systems such as self-organization and minimal models, and subject to the crucial constraint to be in agreement with empirical observations. It is a bit early to say how effective these different models and results will be for urban planning, but the existence of similarities between cities and the convergence between theory and data can help us to see the future of this field with an optimistic eye. Universal results, valid for many cities and independent from their specific evolution, are a good sign for the existence of a single phenomenon – "city" – that is unique and described by a finite set of laws, and that evolves differently because of the specifics of the environment.

References

Acheampong, Ransford Antwi, and Silva, Elisabete. 2015. Land use–transport interaction modeling: A review of the literature and future research directions. *Journal of Transport and Land Use*, **8**(3).

Aldous, David, and Huang, Bowen. 2012. A spatial model of city growth and formation. *arXiv preprint arXiv:1209.5120*.

Aldous, David J., Shun, Julian, et al. 2010. Connected spatial networks over random points and a route-length statistic. *Statistical Science*, **25**(3), 275–288.

Alonso, William, et al. 1964. Location and land use: toward a general theory of land rent. *Location and land use*. Harvard University Press.

Anas, Alex, Arnott, Richard, and Small, Kenneth A. 1998. Urban spatial structure. *Journal of Economic Literature*, 1426–1464.

Anderson, Philip W. 1972. More is different. *Science*, **177**(4047), 393–396.

Antunes, António Lobo, Bavaud, François, and Mager, Christopher. 2009. *Handbook of theoretical and quantitative geography*. Université de Lausanne, Faculté de géosciencies et de l'environment.

Arcaute, Elsa, Hatna, Erez, Ferguson, Peter, Youn, Hyejin, Johansson, Anders, and Batty, Michael. 2013. City boundaries and the universality of scaling laws. *arXiv preprint arXiv:1301.1674*.

Atkinson, Rowland, and Bridge, Gary. 2004. *Gentrification in a global context*. Routledge.

Axhausen, Kay W., and Gärling, Tommy. 1992. Activity-based approaches to travel analysis: conceptual frameworks, models, and research problems. *Transport Reviews*, **12**(4), 323–341.

Baddeley, Alan. 2003. Working memory: looking back and looking forward. *Nature Reviews Neuroscience*, **4**(10), 829–839.

Bak, Per, Tang, Chao, and Wiesenfeld, Kurt. 1987. Self-organized criticality: An explanation of the 1/f noise. *Physical Review Letters*, **59**(4), 381.

Balcan, Duygu, Colizza, Vittoria, Gonçalves, Bruno, Hu, Hao, Ramasco, José J., and Vespignani, Alessandro. 2009. Multiscale mobility networks and the spatial spreading of infectious diseases. *Proceedings of the National Academy of Sciences*, **106**(51), 21484–21489.

Barabasi, Albert-Laszlo. 2005. The origin of bursts and heavy tails in human dynamics. *Nature*, **435**(7039), 207–211.

Barenblatt, Grigory Isaakovich. 1996. *Scaling, self-similarity, and intermediate asymptotics: dimensional analysis and intermediate asymptotics*, vol. 14. Cambridge University Press.

Barthelemy, Marc. 2004. Betweenness centrality in large complex networks. *The European Physical Journal B – Condensed Matter and Complex Systems*, **38**(2), 163–168.

Barthelemy, Marc. 2011. Spatial networks. *Physics Reports*, **499**(1), 1–101.

Barthelemy, Marc, and Flammini, Alessandro. 2008. Modeling urban street patterns. *Physical Review Letters*, **100**(13), 138702.

Barthelemy, Marc, and Flammini, Alessandro. 2009. Co-evolution of density and topology in a simple model of city formation. *Networks and Spatial Economics*, **9**(3), 401–425.

Barthelemy, Marc, Bordin, Patricia, Berestycki, Henri, and Gribaudi, Maurizio. 2013. Self-organization versus top-down planning in the evolution of a city. *Scientific Reports*, **3**, 2153.

Batty, Michael. 2006. Rank clocks. *Nature*, **444**(7119), 592–596.

Batty, Michael. 2007. *Cities and Complexity: Understanding Cities with Cellular Automata, Agent-based Models, and Fractals*. MIT Press.

Batty, Michael. 2008. The size, scale, and shape of cities. *Science*, **319**(5864), 769–771.

Batty, Michael. 2013. *The New Science of Cities*. MIT Press.

Batty, Michael, Carvalho, Rui, Hudson-Smith, Andy, Milton, Richard, Smith, Duncan, and Steadman, Philip. 2008. Scaling and allometry in the building geometries of Greater London. *The European Physical Journal B*, **63**(3), 303–314.

Bazzani, Armando, Giorgini, Bruno, Rambaldi, Sandro, Gallotti, Riccardo, and Giovannini, Luca. 2010. Statistical laws in urban mobility from microscopic GPS data in the area of Florence. *Journal of Statistical Mechanics: Theory and Experiment*, **2010**(05), P05001.

Beckmann, Martin J. 1976. Spatial equilibrium in the dispersed city. In: *Environment, Regional Science and Interregional Modeling*. Springer, pp. 132–141.

Bellman, Richard. 1957. *Dynamic programming*. Princeton University Press.

Benguigui, Lucien. 1992. The fractal dimension of some railway networks. *Journal de Physique I*, **2**(4), 385–388.

Benguigui, Lucien. 1995. A fractal analysis of the public transportation system of Paris. *Environment and Planning A*, **27**(7), 1147–1161.

Bernstein, Jonathan A., Alexis, Neil, Barnes, Charles, Bernstein, I Leonard, Nel, Andre, Peden, David, Diaz-Sanchez, David, Tarlo, Susan M, Williams, P. Brock, and Bernstein, Jonathan A. 2004. Health effects of air pollution. *Journal of Allergy and Clinical Immunology*, **114**(5), 1116–1123.

Bertaud, Alain, and Malpezzi, Stephen. 2003. *The spatial distribution of population in 48 world cities: Implications for economies in transition*. Center for Urban Land Economics Research, University of Wisconsin.

Bettencourt, Luís M.A. 2013. The origins of scaling in cities. *Science*, **340**(6139), 1438–1441.

Bettencourt, Luís M.A., Lobo, José, Helbing, Dirk, Kühnert, Christian, and West, Geoffrey B. 2007. Growth, innovation, scaling, and the pace of life in cities. *Proceedings of the National Academy of Sciences*, **104**(17), 7301–7306.

Bialek, William. 2015. Perspectives on theory at the interface of physics and biology. *arXiv preprint arXiv:1512.08954*.

Black, William R. 1971. An iterative model for generating transportation networks. *Geographical Analysis*, **3**(3), 283288.

Blascheck, T., Kurzhals, K., Raschke, M., Burch, M., Weiskopf, D., and Ertl, T. 2014. State-of-the-art of visualization for eye tracking data. In: *Proceedings of EuroVis*, vol. 2014.

Blondel, Vincent D., Decuyper, Adeline, and Krings, Gautier. 2015. A survey of results on mobile phone datasets analysis. *EPJ Data Science*, **4**(1), 1–55.

Boccaletti, S., Bianconi, G., Criado, R., Del Genio, C.I., Gómez-Gardeñes, J., Romance, M., Sendina-Nadal, I., Wang, Z., and Zanin, M. 2014. The structure and dynamics of multilayer networks. *Physics Reports*, **544**(1), 1–122.

Bouchaud, Jean-Philippe. 2008. Economics needs a scientific revolution. *Nature*, **455**(7217), 1181–1181.

Bouchaud, Jean-Philippe. 2013. Crises and collective socio-economic phenomena: simple models and challenges. *Journal of Statistical Physics*, **151**(3-4), 567–606.

Bouchaud, Jean-Philippe, and Georges, Antoine. 1990. Anomalous diffusion in disordered media: statistical mechanisms, models and physical applications. *Physics Reports*, **195**(4), 127–293.

Bouchaud, Jean-Philippe, and Mézard, Marc. 1997. Universality classes for extreme-value statistics. *Journal of Physics A: Mathematical and General*, **30**(23), 7997.

Bouchaud, Jean-Philippe, and Mézard, Marc. 2000. Wealth condensation in a simple model of economy. *Physica A: Statistical Mechanics and its Applications*, **282**(3), 536–545.

Bouchaud, Jean-Philippe, and Potters, Marc. 2003. *Theory of financial risk and derivative pricing: from statistical physics to risk management*. Cambridge University Press.

Bouttier, Jérémie, Di Francesco, Philippe, and Guitter, Emmanuel. 2004. Planar maps as labeled mobiles. *Electron. J. Combin*, **11**(1), R69.

Bradt, Russell N., Johnson, S.M., and Karlin, Samuel. 1956. On sequential designs for maximizing the sum of n observations. *The Annals of Mathematical Statistics*, 1060–1074.

Branston, David. 1976. Link capacity functions: A review. *Transportation Research*, **10**(4), 223–236.

Bretagnolle, Anne, Pumain, Denise, and Vacchiani-Marcuzzo, Céline. 2009. The organization of urban systems. In: *Complexity perspectives in innovation and social change*. Springer, pp. 197–220.

Bretagnolle, Anne, Delisle, François, et al. 2010. Formes de villes en Europe et aux États-Unis. *Mappemonde*, **97**.

Brockmann, Dirk, Hufnagel, Lars, and Geisel, Theo. 2006. The scaling laws of human travel. *Nature*, **439**(7075), 462–465.

Brueckner, Jan K. 1987. The structure of urban equilibria: A unified treatment of the Muth – Mills model. *Handbook of Regional and Urban Economics*, **2**, 821–845.

Brueckner, Jan K., Thisse, Jacques-Francois, and Zenou, Yves. 1999. Why is central Paris rich and downtown Detroit poor?: An amenity-based theory. *European Economic Review*, **43**(1), 91–107.

Brueckner, Jan K., et al. 2000. Urban sprawl: diagnosis and remedies. *International Regional Science Review*, **23**(2), 160–171.

Brueckner, Jan K., et al. 2011. *Lectures on urban economics*. MIT Press.

Buchanan, Mark. 2014. *Arrogant Physicists – do they think that Economics is easy?*

Calabrese, Francesco, Ferrari, Laura, and Blondel, Vincent D. 2015. Urban sensing using mobile phone network data: a survey of research. *ACM Computing Surveys (CSUR)*, **47**(2), 25.

Cardillo, Alessio, Scellato, Salvatore, Latora, Vito, and Porta, Sergio. 2006. Structural properties of planar graphs of urban street patterns. *Physical Review E*, **73**(6), 066107.

Carra, Giulia, Mulalic, Ismir, Fosgerau, Mogens, and Barthelemy, Marc. 2016. Modeling the relation between commuting distance and income. *Journal of The Royal Society Interface* 13.119 (2016): 20160306.

Chan, S.H.Y., Donner, Reik V., and Lämmer, Stefan. 2011. Urban road networksspatial networks with universal geometric features? *The European Physical Journal B – Condensed Matter and Complex Systems*, **84**(4), 563–577.

Cheng, Zhiyuan, Caverlee, James, Lee, Kyumin, and Sui, Daniel Z. 2011. Exploring millions of footprints in location-sharing services. *ICWSM*, **2011**, 81–88.

Chow, Yuan Shih, Robbins, Herbert, and Siegmund, David. 1971. *Great expectations: The theory of optimal stopping.* Houghton Mifflin.

Christaller, Walter. 1966. *Central places in southern Germany.* Prentice-Hall (English version of the original 1933 paper).

Colizza, Vittoria, Pastor-Satorras, Romualdo, and Vespignani, Alessandro. 2007. Reaction–diffusion processes and metapopulation models in heterogeneous networks. *Nature Physics*, **3**(4), 276–282.

Coniglio, Antonio. 1989. Fractal structure of Ising and Potts clusters: exact results. *Physical Review Letters*, **62**(26), 3054.

Costes, Benoit, Perret, Julien, Bucher, Bénédicte, Gribaudi, Maurizio, An aggregated graph to qualify historical spatial networks using temporal patterns detection. *Proceedings of AGILE, 2015*.

Courtat, Thomas, Gloaguen, Catherine, and Douady, Stephane. 2011. Mathematics and morphogenesis of cities: A geometrical approach. *Physical Review E*, **83**(3), 036106.

Credidio, Heitor F., Teixeira, Elisângela N., Reis, Saulo D.S., Moreira, André A., and Andrade Jr., José S. 2012. Statistical patterns of visual search for hidden objects. *Scientific Reports*, **2**.

Cristelli, Matthieu, Batty, Michael, and Pietronero, Luciano. 2012. There is more than a power law in Zipf. *Scientific Reports*, **2**.

Crucitti, Paolo, Latora, Vito, and Porta, Sergio. 2006. Centrality measures in spatial networks of urban streets. *Physical Review E*, **73**(3), 036125.

Daganzo, Carlos F. 2010. Structure of competitive transit networks. *Transportation Research Part B: Methodological*, **44**(4), 434–446.

Dall'Asta, Luca, Castellano, Claudio, and Marsili, Matteo. 2008. Statistical physics of the Schelling model of segregation. *Journal of Statistical Mechanics: Theory and Experiment*, **2008**(07), L07002.

de Dios Ortúzar, Juan, and Willumsen, Luis G. 1994. *Modelling transport.* John Wiley and Sons.

De Domenico, Manlio, Solé-Ribalta, Albert, Cozzo, Emanuele, Kivelä, Mikko, Moreno, Yamir, Porter, Mason A., Gómez, Sergio, and Arenas, Alex. 2013. Mathematical formulation of multilayer networks. *Physical Review X*, **3**(4), 041022.

Derrible, Sybil, and Kennedy, Christopher. 2009. Network analysis of world subway systems using updated graph theory. *Transportation Research Record: Journal of the Transportation Research Board*, **2112**(-1), 17–25.

Diggle, Peter J., et al. 1983. *Statistical analysis of spatial point patterns.* Academic Press.

Dixon, Philip M. 2002. Ripley's K-function. *Encyclopedia of Environmetrics*, John Wiley and Sons.

Dixon, Philip M., Weiner, Jacob, Mitchell-Olds, Thomas, and Woodley, Robert. 1987. Bootstrapping the Gini coefficient of inequality. *Ecology*, 1548–1551.

Dunbar, Robin I.M. 1992. Neocortex size as a constraint on group size in primates. *Journal of Human Evolution*, **22**(6), 469–493.

Duranton, Gilles, and Diego Puga. 2004. Micro-foundations of urban agglomeration economies. *Handbook of regional and urban economics*, **4**, 2063–2117.

Duplantier, Bertrand. 1989. Statistical mechanics of polymer networks of any topology. *Journal of Statistical Physics*, **54**(3-4), 581–680.

Dupuy, Gabriel, and Benguigui, Lucien Gilles. 2015. Sciences urbaines: interdisciplinarités passive, naïve, transitive, offensive. *Métropoles*.

Dyson, Freeman J. 1962. Statistical theory of the energy levels of complex systems. I. *Journal of Mathematical Physics*, **3**(1), 140–156.

Erlander, Sven, and Stewart, Neil F. 1990. *The gravity model in transportation analysis: theory and extensions*, vol. 3. VSP.

Eubank, Stephen, Guclu, Hasan, Kumar, V.S. Anil, Marathe, Madhav V., Srinivasan, Aravind, Toroczkai, Zoltan, and Wang, Nan. 2004. Modelling disease outbreaks in realistic urban social networks. *Nature*, **429**(6988), 180–184.

Feller, William. 1957. *Introduction to probability theory and its applications*, vol. 1. John Wiley and Sons.

Fialkowski, Marcin, and Bitner, Agnieszka. 2008. Universal rules for fragmentation of land by humans. *Landscape Ecology*, **23**(9), 1013–1022.

Fragkias, Michail, Lobo, José, Strumsky, Deborah, and Seto, Karen C. 2013. Does size matter? Scaling of CO_2 emissions and US urban areas. *PloS One*, **8**(6), e64727.

Freeman, Linton C. 1977. A set of measures of centrality based on betweenness. *Sociometry*, 35–41.

Fujita, Masahisa. 1989. *Urban economic theory: land use and city size*. Cambridge University Press.

Fujita, Masahisa, and Ogawa, Hideaki. 1982. Multiple equilibria and structural transition of non-monocentric urban configurations. *Regional Science and Urban Economics*, **12**(2), 161–196.

Gabaix, Xavier. 1999. Zipf's law for cities: an explanation. *Quarterly Journal of Economics*, 739–767.

Gabaix, Xavier, and Yannis M. Ioannides. 2014. The evolution of city size distributions. Handbook of regional and urban economics 4: 2341–2378.

Gallotti, Riccardo, and Barthelemy, Marc. 2014. Anatomy and efficiency of urban multimodal mobility. *Scientific Reports*, **4**.

Gallotti, Riccardo, and Barthelemy, Marc. 2015. The multilayer temporal network of public transport in Great Britain. *Scientific Data*, **2**.

Gallotti, Riccardo, Armando, Bazzani, and Sandro, Rambaldi. 2012. Towards a statistical physics of human mobility. *International Journal of Modern Physics C*, **23**(09).

Gallotti, Riccardo, Bazzani, Armando, and Rambaldi, Sandro. 2015. Understanding the variability of daily travel-time expenditures using GPS trajectory data. *EPJ Data Science*, **4**(1), 1–14.

Gallotti, Riccardo, Bazzani, Armando, Rambaldi, Sandro, and Barthelemy, Marc. 2016a. A stochastic model of randomly accelerated walkers for human mobility. *Nature Communications* **7** (2016): 12600.

Gallotti, Riccardo, Porter, Mason A., and Barthelemy, Marc. 2016b. Lost in transportation: Information measures and cognitive limits in multilayer navigation. *Science Advances*, **2**(2), e1500445.

Gastner, Michael T., and Newman, Mark E.J. 2006a. Shape and efficiency in spatial distribution networks. *Journal of Statistical Mechanics: Theory and Experiment*, **2006**(01), P01015.

Gastner, Michael T., and Newman, M.E.J. 2006b. Optimal design of spatial distribution networks. *Physical Review E*, **74**(1), 016117.

Gauvin, Laetitia, Vannimenus, Jean, and Nadal, Jean-Pierre, 2009. Phase diagram of a Schelling segregation model. *The European Physical Journal B*, **70**(2), 293–304.

Gibrat, Robert. 1931. *Les inégalités économiques*. Recueil Sirey.

Giuliano, Genevieve, and Small, Kenneth A. 1991. Subcenters in the Los Angeles region. *Regional science and urban economics*, **21**(2), 163–182.

Glaeser, Edward L., and Kahn, Matthew E. 2010. The greenness of cities: carbon dioxide emissions and urban development. *Journal of Urban Economics*, **67**(3), 404–418.

Glaeser, Edward L., Kahn, Matthew E., and Rappaport, Jordan. 2008. Why do the poor live in cities? The role of public transportation. *Journal of Urban Economics*, **63**(1), 1–24.

Glass, L., and Tobler, Waldo R. 1971. Uniform distribution of objects in a homogeneous fields: cities on a plain. *Nature*, **233**, 67–68.

Goldenfeld, Nigel. 1992. *Lectures on phase transitions and the renormalization group*. Addison-Wesley, Advanced Book Program.

Gonçalves, Bruno, Perra, Nicola, and Vespignani, Alessandro. 2011. Modeling users' activity on twitter networks: Validation of Dunbar's number. *PloS One*, **6**(8), e22656.

Gonzalez, Marta C., Hidalgo, Cesar A., and Barabasi, Albert-Laszlo. 2008. Understanding individual human mobility patterns. *Nature*, **453**(7196), 779–782.

Grauwin, Sébastian, Bertin, Eric, Lemoy, Rémi, and Jensen, Pablo. 2009. Competition between collective and individual dynamics. *Proceedings of the National Academy of Sciences*, **106**(49), 20622–20626.

Griffith, Daniel A. 1981. Modeling urban population density in a multi-centered city. *Journal of Urban Economics*, **9**(3), 298–310.

Guérois, Marianne, and Pumain, Denise. 2008. Built-up encroachment and the urban field: a comparison of forty European cities. *Environment and Planning. A*, **40**(9), 2186.

Guibas, Leonidas, and Stolfi, Jorge. 1985. Primitives for the manipulation of general subdivisions and the computation of Voronoi. *ACM Transactions on Graphics (TOG)*, **4**(2), 74–123.

Guo, Zhan, and Wilson, Nigel H.M. 2011. Assessing the cost of transfer inconvenience in public transport systems: A case study of the London Underground. *Transportation Research Part A: Policy and Practice*, **45**(2), 91–104.

Gusein-Zade, Sabir M. 1982. Bunge's problem in central place theory and its generalizations. *Geographical Analysis*, **14**(3), 246–252.

Haggett, Peter, and Chorley, Richard J. 1969. *Network analysis in geography*, vol. 67. Edward Arnold.

Haggett, Peter, Cliff, Andrew David, and Frey, Allan. 1977. Locational analysis in human geography. *Tijdschrift Voor Economische En Sociale Geografie*, **68**(6).

Haggstrom, Gus W. 1966. Optimal stopping and experimental design. *The Annals of Mathematical Statistics*, 7–29.

Hall, Robert E., and Mueller, Andreas I. 2013. Wage dispersion and search behavior. *Unpublished manuscript*.

Hansen, Walter G. 1959. How accessibility shapes land use. *Journal of the American Institute of Planners*, **25**(2), 73–76.

Hawelka, Bartosz, Sitko, Izabela, Beinat, Euro, Sobolevsky, Stanislav, Kazakopoulos, Pavlos, and Ratti, Carlo. 2014. Geo-located Twitter as proxy for global mobility patterns. *Cartography and Geographic Information Science*, **41**(3), 260–271.

Hernando, Alberto, Hernando, R., and Plastino, Alberto, 2014. Space-time correlations in urban sprawl. *Journal of The Royal Society Interface*, **11**(91), 20130930.

Hernando, Alberto, Hernando, R., Plastino, Alberto, and Zambrano, E. 2015. Memory-endowed US cities and their demographic interactions. *Journal of The Royal Society: Interface*, **12**(102), 20141185.

Hilhorst, Hendrik-Jan. 2005. Asymptotic statistics of the n-sided planar Poisson? Voronoi cell: I. Exact results. *Journal of Statistical Mechanics: Theory and Experiment*, **2005**(09), P09005.

Hilhorst, H.J. 2008. Statistical properties of planar Voronoi tessellations. *The European Physical Journal B*, **64**(3-4), 437–441.

Hillier, Bill, and Hanson, Julienne. 1984. *The social logic of space*, vol. 1. Cambridge University Press.

Holme, Petter, and Saramäki, Jari. 2012. Temporal networks. *Physics Reports*, **519**(3), 97–125.

Hornstein, Andreas, Krusell, Per, and Violante, Giovanni L. 2007. *Frictional wage dispersion in search models: A quantitative assessment*. Tech. rept. National Bureau of Economic Research.

Jacobs, Jane. 1961. *The death and life of great American cities*. Vintage.

Jargowsky, Paul A. 1996. Take the money and run: Economic segregation in US metropolitan areas. *American Sociological Review*, 984–998.

Jensen, Pablo. 2006. Network-based predictions of retail store commercial categories and optimal locations. *Physical Review E*, **74**(3), 035101.

Jiang, Bin. 2007. A topological pattern of urban street networks: universality and peculiarity. *Physica A: Statistical Mechanics and its Applications*, **384**(2), 647–655.

Jiang, Bin, and Claramunt, Christophe. 2004. Topological analysis of urban street networks. *Environment and Planning B*, **31**(1), 151–162.

Kang, Chaogui, Ma, Xiujun, Tong, Daoqin, and Liu, Yu. 2012. Intra-urban human mobility patterns: An urban morphology perspective. *Physica A: Statistical Mechanics and its Applications*, **391**(4), 1702–1717.

Kansky, Karel Joseph. 1963. *Structure of transportation networks: relationships between network geometry and regional characteristics*. PhD thesis.

Katifori, Eleni, and Magnasco, Marcelo O. 2012. Quantifying loopy network architectures. *PloS One*, **7**(6), e37994.

Kaufman, L., and Rousseeuw, P.J. 1990. *Finding groups in data*. John Wiley and Sons.

Kivelä, Mikko, Arenas, Alex, Barthelemy, Marc, Gleeson, James P., Moreno, Yamir, and Porter, Mason A. 2014. Multilayer networks. *Journal of Complex Networks*, **2**(3), 203–271.

Krapivsky, Pavel L., Redner, Sidney, and Ben-Naim, Eli. 2010. *A kinetic view of statistical physics*. Cambridge University Press.

Krugman, Paul R. 1996. *The self-organizing economy*. Blackwell.

Lämmer, Stefan, Gehlsen, Björn, and Helbing, Dirk. 2006. Scaling laws in the spatial structure of urban road networks. *Physica A: Statistical Mechanics and its Applications*, **363**(1), 89–95.

Le Néchet, Florent. 2012. Urban spatial structure, daily mobility and energy consumption: A study of 34 European cities. *Cybergeo: European Journal of Geography*, 24966.

Leitao, J.C., Miotto, J.M. Gerlach, M., Altmann, E.G. 2016. Is this scaling nonlinear? R. Soc. open sci. 2016 3 150649.

Lenormand, Maxime, Huet, Sylvie, Gargiulo, Floriana, and Deffuant, Guillaume. 2012. A universal model of commuting networks. *PLoS One*.

Lenormand, Maxime, Picornell, Miguel, Cantú-Ros, Oliva G., Tugores, Antònia, Louail, Thomas, Herranz, Ricardo, Barthelemy, Marc, Frias-Martinez, Enrique, and Ramasco, José J. 2014. Cross-checking different sources of mobility information. *PLoS One*, **9**(8), e105184.

Levinson, David. 2012. Network structure and city size. *PLoS ONE*, **7**(1), e29721.

Levinson, David, and Wu, Yao. 2005. The rational locator reexamined: Are travel times still stable? *Transportation*, **32**(2), 187–202.

Levy, Moshe, and Solomon, Sorin. 1996. Power laws are logarithmic Boltzmann laws. *International Journal of Modern Physics C*, **7**(04), 595–601.

Liang, Xiao, Zheng, Xudong, Lv, Weifeng, Zhu, Tongyu, and Xu, Ke. 2012. The scaling of human mobility by taxis is exponential. *Physica A: Statistical Mechanics and its Applications*, **391**(5), 2135–2144.

Liben-Nowell, David, Novak, Jasmine, Kumar, Ravi, Raghavan, Prabhakar, and Tomkins, Andrew. 2005. Geographic routing in social networks. *Proceedings of the National Academy of Sciences of the United States of America*, **102**(33), 11623–11628.

Lippman, Steven A., and McCall, John. 1976. The economics of job search: A survey. *Economic Inquiry*, **14**(2), 155–189.

Liu, Hsing, Chen, Ying-Hsing, and Lih, Jiann-Shing. 2015. Crossover from exponential to power-law scaling for human mobility pattern in urban, suburban and rural areas. *The European Physical Journal B*, **88**(5), 1–7.

Liu, Yu, Kang, Chaogui, Gao, Song, Xiao, Yu, and Tian, Yuan. 2012. Understanding intra-urban trip patterns from taxi trajectory data. *Journal of Geographical Systems*, **14**(4), 463–483.

Liu, Yu, Sui, Zhengwei, Kang, Chaogui, and Gao, Yong. 2014. Uncovering patterns of inter-urban trip and spatial interaction from social media check-in data. *PloS One*, **9**(1), e86026.

Louail, Thomas, Lenormand, Maxime, Ros, Oliva G. Cantu, Picornell, Miguel, Herranz, Ricardo, Frias-Martinez, Enrique, Ramasco, José J., and Barthelemy, Marc. 2014. From mobile phone data to the spatial structure of cities. *Scientific Reports*, **4**.

Louail, Thomas, Lenormand, Maxime, Picornell, Miguel, Cantú, Oliva García, Herranz, Ricardo, Frias-Martinez, Enrique, Ramasco, José J., and Barthelemy, Marc. 2015. Uncovering the spatial structure of mobility networks. *Nature Communications*, **6**.

Louf, Rémi, and Barthelemy, Marc. 2013. Modeling the polycentric transition of cities. *Physical Review Letters*, **111**(19), 198702.

Louf, Rémi, and Barthelemy, Marc. 2014a. How congestion shapes cities: from mobility patterns to scaling. *Scientific Reports*, **4**.

Louf, Rémi, and Barthelemy, Marc. 2014b. Scaling: lost in the smog. *Environment and Planning B: Planning and Design*, **41**, 767–769.

Louf, Rémi, and Barthelemy, Marc. 2014c. A typology of street patterns. *Journal of The Royal Society Interface*, **11**(101), 20140924.

Louf, Rémi, and Marc Barthelemy. 2016. Patterns of Residential Segregation. PloS one 11.6:e0157476.

Louf, Rémi, Roth, Camille, and Barthelemy, Marc. 2014. Scaling in transportation networks. *PloS One*, **9**(7), e102007.

Luck, Steven J., and Vogel, Edward K. 1997. The capacity of visual working memory for features and conjunctions. *Nature*, **390**(6657), 279–281.

Makse, Hernan A., Havlin, Shlomo, and Stanley, H.E. 1995. Modelling urban growth. *Nature*, **377**, 19.

Malevergne, Yannick, Pisarenko, Vladilen, and Sornette, Didier. 2011. Testing the Pareto against the lognormal distributions with the uniformly most powerful unbiased test applied to the distribution of cities. *Physical Review E*, **83**(3), 036111.

Marchetti, Cesare. 1994. Anthropological invariants in travel behavior. *Technological Forecasting and Social Change*, **47**(1), 75–88.

Marcon, Eric, and Puech, Florence. 2009. Measures of the geographic concentration of industries: improving distance-based methods. *Journal of Economic Geography*, lbp056.

Marshall, Stephen. 2004. *Streets and Patterns*. Routledge.

Massey, Douglas S., and Denton, Nancy A. 1988. The dimensions of residential segregation. *Social Forces*, **67**(2), 281–315.

Masucci, A.P., Smith, D., Crooks, A., and Batty, Michael. 2009. Random planar graphs and the London street network. *The European Physical Journal B-Condensed Matter and Complex Systems*, **71**(2), 259–271.

McCall, John Joseph. 1970. Economics of information and job search. *The Quarterly Journal of Economics*, 113–126.

McMillen, Daniel P. 2001. Nonparametric employment subcenter identification. *Journal of Urban Economics*, **50**(3), 448–473.

McMillen, Daniel P., and Smith, Stefani C. 2003. The number of subcenters in large urban areas. *Journal of Urban Economics*, **53**(3), 321–338.

Mézard, Marc, Parisi, Giorgio, and Virasoro, Miguel-Angel. 1990. *Spin glass theory and beyond*. World Scientific Publishing Co., Inc., Pergamon Press.

Mileyko, Yuriy, Edelsbrunner, Herbert, Price, Charles A., and Weitz, Joshua S. 2012. Hierarchical ordering of reticular networks. *PloS One*, **7**(6), e36715.

Morris, Richard G., and Barthelemy, Marc. 2012. Transport on coupled spatial networks. *Physical Review Letters*, **109**(12), 128703.

Mossay, Pascal, and Picard, Pierre M. 2011. On spatial equilibria in a social interaction model. *Journal of Economic Theory*, **146**(6), 2455–2477.

Muth, Richard F. 1969. *Cities and housing: The spatial pattern of urban residential land use*. University of Chicago Press.

Nadai, Marco De, Staiano, Jacopo, Larcher, Roberto, Sebe, Nicu, Quercia, Daniele, and Lepri, Bruno. 2016. The Death and Life of Great Italian Cities: A Mobile Phone Data Perspective. In: *Proceedings of World Wide Web (WWW)*. ACM.

Newman, Mark. 2000. Applied mathematics: The power of design. *Nature*, **405**(6785), 412–413.

Newman, Mark. 2001. Scientific collaboration networks. II. Shortest paths, weighted networks, and centrality. *Physical Review E*, **64**(1), 016132.

Newman, Peter W.G., and Kenworthy, Jeffrey R. 1989. Gasoline consumption and cities: a comparison of US cities with a global survey. *Journal of the American Planning Association*, **55**(1), 24–37.

Nicosia, Vincenzo, Louf, Rémi, Latora, Vito, and Barthelemy, Marc. 2016. Fragmentation models for road patterns. *Submitted*.

Niedzielski, Michael A., and Malecki, Edward J. 2012. Making tracks: rail networks in world cities. *Annals of the Association of American Geographers*, **102**(6), 1409–1431.

Noulas, Anastasios, Scellato, Salvatore, Lambiotte, Renaud, Pontil, Massimiliano, and Mascolo, Cecilia. 2012. A tale of many cities: universal patterns in human urban mobility. *PloS One*, **7**(5), e37027.

Ogawa, Hideaki, and Fujita, Masahisa. 1980. Equilibrium land use patterns in a nonmonocentric city. *Journal of Regional Science*, **20**(4), 455–475.

Ogawa, Hideaki, and Fujita, Masahisa. 1989. Nonmonocentric urban configurations in a two-dimensional space. *Environment and Planning A*, **21**(3), 363–374.

Okabe, Atsuyuki, and Sadahiro, Y. 1996. An illusion of spatial hierarchy: spatial hierarchy in a random configuration. *Environment and Planning A*, **28**(9), 1533–1552.

Oliveira, Erneson A., Andrade Jr, José S., and Makse, Hernán A. 2014. Large cities are less green. *Scientific Reports*, **4**.

Oreskes, Naomi. 2004. The scientific consensus on climate change. *Science*, **306**(5702), 1686–1686.

O'Sullivan, David, and Manson, Steven M. 2015. Do physicists have geography envy? And what can geographers learn from it? *Annals of the Association of American Geographers*, **105**(4), 704–722.

Paas, Fred, Renkl, Alexander, and Sweller, John. 2003. Cognitive load theory and instructional design: Recent developments. *Educational Psychologist*, **38**(1), 1–4.

Pan, Wei, Ghoshal, Gourab, Krumme, Coco, Cebrian, Manuel, and Pentland, Alex. 2013. Urban characteristics attributable to density-driven tie formation. *Nature Communications*, **4**.

Penn, Alan. 2003. Space syntax and spatial cognition or why the axial line? *Environment and Behavior*, **35**(1), 30–65.

Perret, Julien, Gribaudi, Maurizio, and Barthelemy, Marc. 2015. Roads and cities of 18th-century France. *Scientific Data*, **2**.

Porta, Sergio, Crucitti, Paolo, and Latora, Vito. 2006a. The network analysis of urban streets: A dual approach. *Physica A: Statistical Mechanics and its Applications*, **369**(2), 853–866.

Porta, Sergio, Crucitti, Paolo, and Latora, Vito. 2006b. The network analysis of urban streets: A primal approach. *Environment and Planning B: Planning and design*, **33**, 705–725.

Porta, Sergio, Strano, Emanuele, Iacoviello, Valentino, Messora, Roberto, Latora, Vito, Cardillo, Alessio, Wang, Fahui, and Scellato, Salvatore. 2009. Street centrality and densities of retail and services in Bologna, Italy. *Environment and Planning B: Planning and design*, **36**(3), 450–465.

Pumain, Denise. 1997. Pour une théorie évolutive des villes. *Espace géographique*, **26**(2), 119–134.

Pumain, Denise. 2004. Scaling laws and urban systems. *Santa Fe Institute, Working Paper n 04-02*, **2**, 26.

Pumain, Denise and Moriconi-Ebrard, F. 1997. City size distribution and metropolisation. *Geojournal*, **43**, 307–314.

Rasmussen, Shelley L., and Starr, Norman. 1979. Optimal and adaptive stopping in the search for new species. *Journal of the American Statistical Association*, **74**(367), 661–667.

Reed, William J., and Hughes, Barry D. 2002. From gene families and genera to incomes and internet file sizes: Why power laws are so common in nature. *Physical Review E*, **66**(6), 067103.

Reynolds-Feighan, Aisling. 2001. Traffic distribution in low-cost and full-service carrier networks in the US air transportation market. *Journal of Air Transport Management*, **7**(5), 265–275.

Rhee, Injong, Shin, Minsu, Hong, Seongik, Lee, Kyunghan, Kim, Seong Joon, and Chong, Song. 2011. On the levy-walk nature of human mobility. *IEEE/ACM transactions on networking (TON)*, **19**(3), 630–643.

Rosvall, Martin, Trusina, Ala, Minnhagen, Petter, and Sneppen, Kim. 2005. Networks and cities: An information perspective. *Physical Review Letters*, **94**(2), 028701.

Roth, Camille, Kang, Soong Moon, Batty, Michael, and Barthelemy, Marc. 2011. Structure of urban movements: polycentric activity and entangled hierarchical flows. *PloS One*, **6**(1), e15923.

Roth, Camille, Kang, Soong Moon, Batty, Michael, and Barthelemy, Marc. 2012. A long-time limit for world subway networks. *Journal of The Royal Society Interface*, **9**(75), 25402550.

Rozenblat, Céline, and Pumain, Denise. 2007. Firm linkages, innovation and the evolution of urban systems. *Cities in Globalization: Practices, Policies, and Theories*, 130–156.

Rozenfeld, Hernán D., Rybski, Diego, Andrade, José S., Batty, Michael, Stanley, H. Eugene, and Makse, Hernán A. 2008. Laws of population growth. *Proceedings of the National Academy of Sciences*, **105**(48), 18702–18707.

Rozenfeld, Hernán D., Diego Rybski, Xavier Gabaix, and Hernn A. Makse. 2011. The area and population of cities: New insights from a different perspective on cities. The American Economic Review 101, no. 5: 2205–2225.

Runions, Adam, Fuhrer, Martin, Lane, Brendan, Federl, Pavol, Rolland-Lagan, Anne-Gaëlle, and Prusinkiewicz, Przemyslaw. 2005. Modeling and visualization of leaf venation patterns. *ACM Transactions on Graphics (TOG)*, vol. 24. ACM, pp. 702–711.

Rybski, Diego, Dominik E. Reusser, Anna-Lena Winz, Christina Fichtner, Till Sterzel, and Jrgen P. Kropp. 2016. Cities as nuclei of sustainability?. Environment and Planning B: Planning and Design: 0265813516638340.

Samaniego, Horacio, and Moses, Melanie E. 2008. Cities as organisms: Allometric scaling of urban road networks. *Journal of Transport and Land Use*, **1**(1).

Schelling, Thomas C. 1971. Dynamic models of segregation. *Journal of Mathematical Sociology*, **1**(2), 143–186.

Schläpfer, Markus, Lee, Joey, and Bettencourt, Luís. 2015. Urban Skylines: Building heights and shapes as measures of city size. *arXiv preprint arXiv:1512.00946*.

Schwarz, Nina. 2010. Urban form revisited: Selecting indicators for characterising European cities. *Landscape and Urban Planning*, **96**(1), 29–47.

Seidman, Stephen B. 1983. Network structure and minimum degree. *Social Networks*, **5**(3), 269–287.

Sen, Parongama. 1999. Nonlocal conservation in the coupling field: effect on critical dynamics. *Journal of Physics A: Mathematical and General*, **32**(9), 1623.

Shiryaev, Albert N. 1963. On optimum methods in quickest detection problems. *Theory of Probability & Its Applications*, **8**(1), 22–46.

Simini, Filippo, González, Marta C., Maritan, Amos, and Barabási, Albert-László. 2012. A universal model for mobility and migration patterns. *Nature*, **484**(7392), 96–100.

Song, Chaoming, Qu, Zehui, Blumm, Nicholas, and Barabási, Albert-László. 2010a. Limits of predictability in human mobility. *Science*, **327**(5968), 1018–1021.

Song, Chaoming, Koren, Tal, Wang, Pu, and Barabási, Albert-László. 2010b. Modelling the scaling properties of human mobility. *Nature Physics*, **6**(10), 818–823.

Soo, Kwok Tong. 2005. Zipf's Law for cities: a cross-country investigation. *Regional Science and Urban Economics*, **35**(3), 239–263.

Sornette, Didier, and Cont, Rama. 1997. Convergent multiplicative processes repelled from zero: power laws and truncated power laws. *Journal de Physique I*, **7**(3), 431–444.

Southworth, Michael, and Ben-Joseph, Eran. 2003. *Streets and the Shaping of Towns and Cities*. Island Press.

Stanley, H. Eugene. 1971. *Introduction to phase transitions and critical phenomena*. Oxford University Press.

Stigler, George J. 1961. The economics of information. *The Journal of Political Economy*, 213–225.

Strano, Emanuele, Nicosia, Vincenzo, Latora, Vito, Porta, Sergio, and Barthelemy, Marc. 2012. Elementary processes governing the evolution of road networks. *Scientific Reports*, **2**.

Strano, Emanuele, Shai, Saray, Dobson, Simon, and Barthelemy, Marc. 2015. Multiplex networks in metropolitan areas: generic features and local effects. *Journal of The Royal Society Interface*, **12**(111), 20150651.

Sung, Hyungun, Lee, Sugie, and Cheon, SangHyun. 2015. Operationalizing Jane Jacobs's urban design theory: empirical verification from the great city of Seoul, Korea. *Journal of Planning Education and Research*, 0739456X14568021.

Tang, Jinjun, Liu, Fang, Wang, Yinhai, and Wang, Hua. 2015. Uncovering urban human mobility from large-scale taxi GPS data. *Physica A: Statistical Mechanics and its Applications*, **438**, 140–153.

Thisse, Jacques-François. 2014. The New Science of Cities by Michael Batty: The opinion of an economist. *Journal of Economic Literature*, **52**(3), 805–819.

Tsai, Yu-Hsin. 2005. Quantifying urban form: compactness versus sprawl. *Urban Studies*, **42**(1), 141–161.

Venerandi, Alessandro, Zanella, Mattia, Romice, Ombretta, and Porta, Sergio. 2014. The Form of Gentrification. *arXiv preprint arXiv:1411.2984*.

Viana, Matheus P., Strano, Emanuele, Bordin, Patricia, and Barthelemy, Marc. 2013. The simplicity of planar networks. *Scientific Reports*, **3**.

Vinković, Dejan, and Kirman, Alan. 2006. A physical analogue of the Schelling model. *Proceedings of the National Academy of Sciences*, **103**(51), 19261–19265.

Von Thunen, Johann Heinrich, and Hall, Peter Geoffrey. 1966. *Isolated state*. Pergamon.

Wald, A. 1947. *Sequential Analysis. 1947*.

Wang, Pu, Hunter, Timothy, Bayen, Alexandre M., Schechtner, Katja, and González, Marta C. 2012. Understanding road usage patterns in urban areas. *Scientific Reports*, **2**.

Wang, Wenjun, Pan, Lin, Yuan, Ning, Zhang, Sen, and Liu, Dong. 2015. A comparative analysis of intra-city human mobility by taxi. *Physica A: Statistical Mechanics and its Applications*, **420**, 134–147.

Weiner, Edward. 1999. *Urban transportation planning in the United States: An historical overview*. Greenwood Publishing Group.

Wilkerson, Galen, Khalili, Ramin, and Schmid, Stefan. 2013. Urban mobility scaling: Lessons from little data. *arXiv preprint arXiv:1401.0207*.

Wilson, Alan Geoffrey. 1969. The use of entropy maximising models, in the theory of trip distribution, mode split and route split. *Journal of Transport Economics and Policy*, 108–126.

Xie, Feng, and Levinson, David. 2007. Measuring the structure of road networks. *Geographical Analysis*, **39**(3), 336–356.

Xie, Feng, and Levinson, David M. 2011. *Evolving transportation networks*, vol. 1. Springer.

Yan, Xiao-Yong, Han, Xiao-Pu, Wang, Bing-Hong, and Zhou, Tao. 2013. Diversity of individual mobility patterns and emergence of aggregated scaling laws. *Scientific Reports*, **3**.

Yerra, Bhanu M., and Levinson, David M. 2005. The emergence of hierarchy in transportation networks. *The Annals of Regional Science*, **39**(3), 541–553.

Zaburdaev, V., Denisov, S., and Klafter, J. 2015. Lévy walks. *Reviews of Modern Physics*, **87**(2), 483.

Zahavi, Yacov. 1974. *Traveltime budgets and mobility in urban areas*. Tech. rept.

Zhao, Kai, Musolesi, Mirco, Hui, Pan, Rao, Weixiong, and Tarkoma, Sasu. 2015. Explaining the power-law distribution of human mobility through transportation modality decomposition. *Scientific Reports*, **5**.

Zheng, Yu, Capra, Licia, Wolfson, Ouri, and Yang, Hai. 2014. Urban computing: Concepts, methodologies, and applications. *ACM Transactions on Intelligent Systems and Technology (TIST)*, **5**(3), 38.

Zipf, George Kingsley. 1946. The P1 P2/D hypothesis: On the intercity movement of persons. *American Sociological Review*, 677–686.

Zipf, George Kingsley. 1949. *Human behavior and the principle of least effort.* Addison-Wesley Press.

Index

CO_2 emissions, 72, 74

absentee landlords, 199
accessibility, 107
Alonso–Muth–Mills model, 193
AMM model, 193, 194
area of cities, 21
area versus population, 70

Base Transceiver Station, 130
Beckmann's model, 200
Bellman equation, 145
betweenness centrality, 89, 174
betweenness centrality in a multilayer network, 183
betweenness centrality, spatial distribution of, 183
bid-rent function, 197
bits, 187
block, 81
branches and core structure, 117
building, 18
building volume, 18

CDR, 130
Central Business District, 3
central place theory, 236
City Clustering Algortithm, 5
closed city, 198
closest opportunity model, 145, 147
cognitive threshold, 190
collective advantage, 218
commuting function, 58
commuting networks, 132
composite commidities, 34
cooperation, 218
cooperative dynamics, 217
counting hotspots, 56
coupling, 165
coverage, 121

definition of cities, 5
diffusion with noise, 228, 231
digital revolution, 246
dimensional analysis, 221
dynamics of segregation, 214

edge-city model, 63
effective acceleration, 155
emergent behavior, 25
entropy, 187
equilibrium, 32
European versus US cities, 4
exposure, 206

footprint area, 18
fractals, 117
fragmentation models, 109
Fujita–Ogawa model, 57

gasoline consumption, 72
generalized cost, 210
gentrification, 203
Gini coefficient, 174
GIS, 40
gravity model, 138

hotspots, 47, 52

identifying hotspots, 54
income elasticity, 212
indifference curve, 197
indirect utility function, 197
information entropy, 186
information threshold, 188
interactions, 32
interdependence profile, 180
Interdependency, 165
interdependency of a node, 179
interdisciplinarity, 45

local maximum of density, 52
local environment quality, 51
locational potential, 58
Lorenz curve, 54
Loubar method, 54
LUTI models, 29

M index, 50
Marchetti's constant, 19
market potential, 63
minimal path, 166
Monocentric structure, 3
MSA, 5
multilayer networks, 165
multiplex, 173

naïve scaling, 20
Nash equilibrium, 215
number of cities, 226
number of hotspots, 68

OD matrix, 130
optimal strategy, 145
origin–destination matrix, 130
outreach, 181

percolation transition, 6
polycentric structure, 24
primacy index, 227

quality, 139
quickest path, 166

radiation model, 138
rank ordered plot, 12
rational locator hypothesis, 19
relevant parameters, 28
remote working, 246
renormalization group, 28
representation, 205
resolution parameter, 52
retail stores, 49
road network, 22

scaling, 219, 220
scaling exponents, 220

Schelling model, 214, 217
segregation, 203
simplest paths, 92, 186
simplicity index, 92, 93
simplicity profile, 93
social classes, 203
social interactions, 200
social optimum, 217
spatial correlations, 234
spatial dominance, 238
spatial hierarchy, 238
spatial networks, 78
steady state, 120
steepest gradient, 211
stochastic diffusion, 228
stop events frequency, 169
street network, 22, 78
streetcar suburbs, 3
stylized fact, 29
subway branches, 117
subways, 4
sustainability, 74
synchronization effect, 168
synchronization inefficiency, 167
systems of cities, 225

temporal correlations, 234
time-respecting path, 166
total commuting distance, 23, 71
total delay due to congestion, 72
total length of roads, 21, 71
transportation systems, 165
trip anatomy, 164

universality, 28
urban computing, 245
urban horizon, 181
urban village, 75
utility, 214

value of time, 34, 210
Voronoi cell, 48, 238
Voronoi tessellation, 48, 238

Zipf plot, 12